Artificial Intelligence and Ambient Intelligence

Artificial Intelligence and Ambient Intelligence

Editors

Matjaz Gams
Martin Gjoreski

MDPI • Basel • Beijing • Wuhan • Barcelona • Belgrade • Manchester • Tokyo • Cluj • Tianjin

Editors
Matjaz Gams
Department of
intelligent systems
Jozef Stefan Institute
Ljubljana
Slovenia

Martin Gjoreski
Computer Science
Università della Svizzera Italiana
Lugano
Switzerland

Editorial Office
MDPI
St. Alban-Anlage 66
4052 Basel, Switzerland

This is a reprint of articles from the Special Issue published online in the open access journal *Electronics* (ISSN 2079-9292) (available at: www.mdpi.com/journal/electronics/special_issues/AIs_electronics).

For citation purposes, cite each article independently as indicated on the article page online and as indicated below:

LastName, A.A.; LastName, B.B.; LastName, C.C. Article Title. *Journal Name* **Year**, *Volume Number*, Page Range.

ISBN 978-3-0365-2533-4 (Hbk)
ISBN 978-3-0365-2532-7 (PDF)

© 2021 by the authors. Articles in this book are Open Access and distributed under the Creative Commons Attribution (CC BY) license, which allows users to download, copy and build upon published articles, as long as the author and publisher are properly credited, which ensures maximum dissemination and a wider impact of our publications.

The book as a whole is distributed by MDPI under the terms and conditions of the Creative Commons license CC BY-NC-ND.

Contents

About the Editors vii

Preface to "Artificial Intelligence and Ambient Intelligence" ix

Matjaz Gams and Martin Gjoreski
Artificial Intelligence and Ambient Intelligence
Reprinted from: *Electronics* **2021**, *10*, 941, doi:10.3390/electronics10080941 1

Matjaž Gams and Tine Kolenik
Relations between Electronics, Artificial Intelligence and Information Society through Information Society Rules
Reprinted from: *Electronics* **2021**, *10*, 514, doi:10.3390/electronics10040514 7

Jaakko Tervonen, Kati Pettersson and Jani Mäntyjärvi
Ultra-Short Window Length and Feature Importance Analysis for Cognitive Load Detection from Wearable Sensors
Reprinted from: *Electronics* **2021**, *10*, 613, doi:10.3390/electronics10050613 23

Tomaž Kompara, Janez Perš, David Susič and Matjaž Gams
A One-Dimensional Non-Intrusive and Privacy-Preserving Identification System for Households
Reprinted from: *Electronics* **2021**, *10*, 559, doi:10.3390/electronics10050559 43

Ramon F. Brena, Edgar Escudero, Cesar Vargas-Rosales, Carlos E. Galvan-Tejada and David Munoz
Device-Free Crowd Counting Using Multi-Link Wi-Fi CSI Descriptors in Doppler Spectrum
Reprinted from: *Electronics* **2021**, *10*, 315, doi:10.3390/electronics10030315 65

Hao-Chiang Koong Lin, Yu-Chun Ma and Min Lee
Constructing Emotional Machines: A Case of a Smartphone-Based Emotion System
Reprinted from: *Electronics* **2021**, *10*, 306, doi:10.3390/electronics10030306 91

Michal Bednarek, Piotr Kicki, Jakub Bednarek and Krzysztof Walas
Gaining a Sense of Touch Object Stiffness Estimation Using a Soft Gripper and Neural Networks
Reprinted from: *Electronics* **2021**, *10*, 96, doi:10.3390/electronics10010096 111

Tomasz Mańkowski, Jakub Tomczyński, Krzysztof Walas and Dominik Belter
PUT-Hand—Hybrid Industrial and Biomimetic Gripper for Elastic Object Manipulation
Reprinted from: *Electronics* **2020**, *9*, 1147, doi:10.3390/electronics9071147 127

Michal Bednarek, Piotr Kicki and Krzysztof Walas
On Robustness of Multi-Modal Fusion—Robotics Perspective
Reprinted from: *Electronics* **2020**, *9*, 1152, doi:10.3390/electronics9071152 153

About the Editors

Matjaz Gams

Matjaz Gams is Head of the department of intelligent systems at the Jozef Stefan Institute and professor of computer science at the University of Ljubljana and MPS, Slovenia. He is or was teaching at 10 Faculties in Slovenia and Germany. His professional interests include information society, intelligent systems, artificial intelligence and cognitive. He is managing director of the *Informatica* journal, published for 45 years. He has been president of the Organizing Committee of the IS conference for 24 years, delivering the prestigious Michie-Turing award for life achievements. He was co-founder of various societies in Slovenia, e.g., the Engineering Academy, AI Society, Cognitive Society, ACM Slovenia, SLAIS. He cooperated/headed around two hundred national and international projects including top EU projects FP and H2020. His team was awarded with several research and innovation top national and international achievements including second place at the XPRIZE COVID-19 countermeasures competition. His publication list includes 150 original scientific publications and 1500 items in all categories, including 7 patent applications.

Martin Gjoreski

Martin Gjoreski is a postdoc researcher at the Faculty of Informatics, Università della Svizzera italiana (USI), in the research group of prof. Marc Langheinrich and Prof. Silvia Santini. His research field is Artificial Intelligence with machine learning and wearable computing. He did his doctoral as part of the Department of Intelligent Systems at the Jozef Stefan Institute advised by Prof. Dr. Matjaž Gams and Dr. Mitja Luštrek.

For his PhD thesis, he was awarded the Slovenian prize "Jožef Stefan Golden Emblem" for outstanding PhD thesis. His research work has been presented at dozens of international conferences and published in several international journals. He was also part of the teams that won five machine-learning challenges organized at A-ranked conferences (UbiComp, and ABC Conference): SHL 2018, SHL 2019, EmteqAR 2019, Challenge-UP 2019 and Cooking AR Challenge 2020.

Preface to "Artificial Intelligence and Ambient Intelligence"

This book includes a series of scientific papers published in the Special Issue on Artificial Intelligence and Ambient Intelligence at the journal *Electronics* MDPI. The book starts with an opinion paper on "Relations between Electronics, Artificial Intelligence and Information Society through Information Society Rules", presenting relations between information society, electronics and artificial intelligence mainly through twenty-four IS laws. After that, the book continues with a series of technical papers which present applications of Artificial Intelligence and Ambient Intelligence in a variety of fields including affective computing, privacy and security in smart environments, and robotics. More specifically, the first part presents usage of Artificial Intelligence (AI) methods in combination with wearable devices (e.g., smartphones and wristbands) for recognizing human psychological states (e.g., emotions and cognitive load). The second part presents usage of AI methods in combination with laser sensors or Wi-Fi signals for improving security in smart buildings by identifying and counting the number of visitors. The last part presents usage of AI methods in robotics for improving robots' ability for object gripping manipulation and perception. The language of the book is rather technical, thus the intended audience are scientists and researchers who have at least some basic knowledge in computer science.

Matjaz Gams, Martin Gjoreski
Editors

Editorial

Artificial Intelligence and Ambient Intelligence

Matjaz Gams *[ID] and Martin Gjoreski [ID]

Department of Intelligent Systems, Jozef Stefan Institute, Jamova 39, 1000 Ljubljana, Slovenia; martin.gjoreski@ijs.si
* Correspondence: matjaz.gams@ijs.si

1. Introduction

Artificial intelligence (AI) and its sister ambient intelligence (AmI) have in recent years become one of the main contributors to the progress of digital society and human civilization. For example, breakthroughs have been achieved in image processing [1–4] natural language processing [5–7], and reinforcement learning [8,9]. All of this affects practically every aspect of our lives, be it search engines such as Google, autonomous vehicles, robots, or smart healthcare. The relation to electronics is particularly interesting. While the exponential progress of electronics expressed through Moore's Law [10] or Keck's Law enabled progress of information society and AI, the design of new chips already to some extent depends on the successful application of AI methods, and will likely more so in the future.

Several questions arise in relation to the above research and development fields. Are there major possibilities for improvements by connecting SW, AI, and AmI methods directly to the chips? Is it possible to integrate the flexibility of SW with the speed of electronic HW and vastly improve the cognitive and computing powers? Will AmI benefit through this progress, since it is intrinsically devoted to connecting devices and humans?

However, future is all but certain as the COVID-19 crisis demonstrates. It might be that we are already facing a slow but steady decline of electronic components following the fast exponential growth. In addition, AI is notoriously known for its wild ups and downs similar to computer generations, where after a hype a major disappointment is proclaimed worldwide when the human level intelligence seems to be as far as before [11]. However, like Phoenix, AI rises again and again, and unlike well-known physical hardware limitations there is no major well-defined limitation for the AI progress. Indeed, it seems that superintelligence and super ambient intelligence are just decades away [12]. They will bring major technological and societal changes, hopefully for the best.

Citation: Gams, M.; Gjoreski, M. Artificial Intelligence and Ambient Intelligence. *Electronics* **2021**, *10*, 941. https://doi.org/10.3390/electronics10080941

Received: 8 April 2021
Accepted: 9 April 2021
Published: 15 April 2021

Publisher's Note: MDPI stays neutral with regard to jurisdictional claims in published maps and institutional affiliations.

The objective of this Special Issue is to focus on the technical and overview contribution for the AI, AmI, information society and electronics. In addition, papers deal with

- Mobile/wearable intelligence
- Robotics applied to smart tasks
- Applications of combined pervasive/ubiquitous/cognitive computing with AI
- Use of mobile, wireless, visual, and multi-modal sensor networks in intelligent systems
- Intelligent handling of privacy, security and trust

2. Artificial Intelligence and Ambient Intelligence

Copyright: © 2021 by the authors. Licensee MDPI, Basel, Switzerland. This article is an open access article distributed under the terms and conditions of the Creative Commons Attribution (CC BY) license (https://creativecommons.org/licenses/by/4.0/).

In the review paper "Relations between Electronics, Artificial Intelligence and Information Society through Information Society Rules" [13], Matjaž Gams at al. present relations between information society (IS), electronics and artificial intelligence mainly through twenty-four IS laws. The laws constitute a novel collection, not presented in literature before, describing major properties in the mentioned field, and the way they influence progress. The laws mainly describe the exponential growth in a particular field such as processing, storage or transmission capabilities with related references for further study.

Each law bears the name of its inventor. Rules such as Moore's Law are reasonably well known even in general public, however, the majority of rules is not presented at university education all over the world. There exist probably tens of similar rules, but the authors picked the most relevant to comprehensibly present the fields. Not all rules are technical, some present relations to production prices and human interaction while others capture human cognitive issues. An analysis is devoted to time dependencies of the rules, and the final part of the paper describes the progress, state-of the-art and potential further progress of AI. AI is already occasionally exceeding human capabilities and will do so even more in the future. In some areas where AI was presumed to be incapable of performing even at a modest level, such as the production of art or programming software, AI is making progress that can sometimes reflect true human skills by programs like GPT3.

The review paper is followed by seven research papers.

Jaakko Tervonen et al. [14] addressed the issue of human cognitive abilities under pressure in the information society in "Ultra-Short Window Length and Feature Importance Analysis for Cognitive Load Detection from Wearable Sensors". Cognitive load detection is beneficial in several applications of human–computer interaction, for example in autonomous driving. The paper concentrates on accurate and real-time bio signal-based cognitive-load detection. More specifically, the paper addresses the problem of data segmentation by analyzing optimal and minimal window length. A comparative analysis is presented, in which ultra-short (30 s or less) window lengths were used for cognitive load detection with a wrist-worn device, which provides heart rate, heart rate variability, galvanic skin response, and skin temperature. These bio signal data are used to extract features at six different window lengths. The extracted features are then used to train an Extreme Gradient Boosting classifier to detect high vs. low cognitive load. The results indicate that longer intervals in general achieve higher accuracy, with 25 s window performing the best (67.6%). Lowest performance (60.0%) is obtained with 5 s window. The relation between different bio signal features, the classification performance and the most useful features was also investigated. The results with wearables seem as reliable as with other, more expensive and obtrusive sensors.

The article "A One-Dimensional Non-Intrusive and Privacy-Preserving Identification System for Households" by Tomaz Kompara et al. [15] introduces a novel indoor identification system based on a network of laser sensors, each attached on top of the room entry. There is a need for systems awareness of an inhabitant's presence and identity in many ambient-intelligence applications, including intelligent homes and cities, with two major concerns: costs and preserving non-intrusiveness. The system should be seamless for the user, preserving the user's privacy as much as possible. The proposed solution is based on a one-dimensional depth sensor, mounted on top of a doorway, facing towards the entrance at an angle. This position allows acquiring the user's body shape, i.e., silhouette, while the user is crossing the doorway. The sensor data coupled with classical machine learning methods are used for user-identification. The system is non-intrusive and preserves privacy. This is achieved by omitting user-sensitive information such as activity, facial expression or clothing. Additionally, the system does not use video or audio data. The system is based on a statistical observation that a typical household is shared by only a small number of physically quite different inhabitants. This hypothesis was tested on a nearly 4000-person, publicly available database of anthropometric measurements. The analysis of the relationships among accuracy, measured data and number of residents revealed quality accuracy up to 10 inhabitants. In addition, the system was evaluated in a real-world scenario on 18 subjects entering a door under a variety of conditions (e.g., different objects and different clothing). A 10-fold cross validation showed 98.4% accuracy for all subjects, and 99.1% for groups of five subjects. These results indicate that a network of one-dimensional depth sensors might be suitable for the identification task with purposes such as non-obtrusive surveillance for security and ambient-intelligence comfort.

In "Device-Free Crowd Counting Using Multi-Link Wi-Fi CSI Descriptors in Doppler Spectrum" [16], Ramon F. Brena et al., tasked themselves to successfully measure the quantity of people in a given space. This information is relevant in many applications, ranging from marketing to safety. The approach is based on measuring crowd size with an inexpensive Wi-Fi equipment, taking advantage of the fact that Wi-Fi signals get distorted by people's presence. Based on the previous experience and by identifying distortion Wi-Fi patterns, the method estimates the number of people in a given space. Using machine learning classifiers and channel state information (CSI), the method estimates the number of people placed between a Wi-Fi transmitter and a receiver. The method achieved better results than the compared single link or averaging approaches. The advantage comes from taking into consideration individual channel information instead of taking the average of the information of all channels. The experiments demonstrated improvements from 44% accuracy with one link to 99% with six links. Additionally, more details are presented about how the addition of each of the multiple links of information influences the accuracy of the prediction.

In "Constructing Emotional Machines: A Case of a Smartphone-Based Emotion System" by Hao-Chiang Koong Lin et al. [17], the emphasis is on an emotion system (emotion machines) developed and deployed on smartphones. The objective of this study is to explore factors that developers focus on when developing emotional machines. More specifically, user attitudes toward emotional messages sent by machines and the effects of emotion systems on user behavior were investigated in detail. A study was performed for two weeks with 124 individuals using a smartphone for more than one year. The participants used the system at will and freely interacted with the system agent. The smartphones generated 11,264 crucial notifications in total, among which 76% were viewed by the participants and 68.1% enabled the participants to resolve unfavorable smartphone conditions in a timely manner and allowed the system agent to provide users with positive emotional feedback. The majority of the participants were pleased by the emotional messages, they were taking into account the emotional messages and were convinced that the developed system enabled their smartphone to exhibit emotions. Additionally, a study revealed that an emotion system triggers certain patterns and behaviors in users, and the degree of attention paid to emotional messages corresponds to the quality of the emotion system.

In "Gaining a Sense of Touch Object Stiffness Estimation Using a Soft Gripper and Neural Networks" [18], Michal Bednarek et al. deal with soft gripping. The objective is to manipulate an elastic, soft and unstructured object, vulnerable to deformations. To perform such a task successfully, it is necessary to estimate the physical parameters of a squeezed object to adjust the manipulation procedure. While humans perform the task using a large volume of obtained knowledge starting from childhood, robots lack that type of knowledge and must rely on other approaches. The chosen approach is based on estimation of physical parameters using deep learning algorithms utilizing measurements from direct interaction with objects using robotic grippers. The interaction of the gripper with the object generates signals which are used to calculate object stiffness coefficient. Physical experiments were executed by the Yale OpenHand soft gripper, based on readings from inertial measurement units (IMUs) attached to the fingers of the gripper. The results indicate that the approach can reliably estimate the parameters of the object thus enabling smooth grasping and handling. The results enabled the creation of three datasets of IMU readings gathered while squeezing the objects, two from the experiments in simulation environment and one from real-life experiments. The dataset is publicly available to the scientific community to enable further testing of new approaches in the growing field of soft manipulation.

The paper "On Robustness of Multi-Modal Fusion—Robotics Perspective" [19] by Michal Bednarek et al. deals with a robotic perception system that needs to successfully integrate information from several data streams. Multi-modal fusion of heterogeneous data streams is a crucial ability enabling noise-robustness. Related approaches often rely on application-specific manual design of a multimodal-data fusion system to handle multi-modal data. As the volume and dimensionality of sensory feedback increase in recent years, it is beneficial to use other approaches. Multi-modal machine learning is one of the emerging fields for this task with focus mainly on vision and audio input. Robots, however, often use haptic sensors when interacting with an environment. An example would be gripping an object and handling it in a particular way. The experiments described in the paper involved three tasks: (i) grasp outcome classification, (ii) texture recognition, and (iii) multi-label classification of haptic adjectives based on haptic and visual data. Four learning-based multi-modal fusion methods were compared on three publicly available datasets containing haptic signals, images, and robots' poses. The quality of each method was analyzed, in terms of performing the task and on their robustness against data degradation. The later issue is rarely considered in the research papers, whereas it is quite common in real life, when a degradation of sensory feedback often occurs during robot interaction with its environment, e.g., under various light conditions.

In "PUT-Hand—Hybrid Industrial and Biomimetic Gripper for Elastic Object Manipulation" [20], Tomasz Mańkowski et al. present an approach for manipulation of elastic objects using an anthropomorphic gripper based on off-the-shelf and 3D-printed components. The gripper contains five elements and each of them contains three fully actuated fingers for precise manipulation, and two tendon-driven digits for secure power grasping. The gripper is equipped with an on-board controller circuit and firmware, enabling full joint control and observation by resistive position and angle sensors in each joint. Additionally, the sensory system of the hand consists of tri-axial optical force sensors placed on fully actuated fingers' fingertips for reaction force measurement. A PC provides the motor control using USB communication protocol providing a robot operating system in the form of a driver. To analyze performance of the gripper, several experiments were performed and are reported in the paper. The design files, source codes and results are available online under CC BY-NC 4.0 and MIT licenses.

3. Conclusions

We would like to take this opportunity to thank all the authors for submitting papers to this Special Issue. We also hope that the readers will find new and useful information on artificial intelligence and ambient intelligence as this field continues to progress with amazing speed.

Acknowledgments: We would like to thank all the researchers who submitted articles to this special issue. We are also grateful to all the reviewers who helped in the evaluation of the manuscripts and made very valuable suggestions to improve the quality of contributions. We would like to acknowledge the editorial board of Electronics, who invited us to guest edit this special issue. We are also grateful to the Electronics Editorial Office staff who worked thoroughly to maintain the rigorous peer-review schedule and timely publication.

Conflicts of Interest: The authors declare no conflict of interest.

References

1. Krizhevsky, A.; Sutskever, I.; Hinton, G.E. Imagenet classification with deep convolutional neural networks. *Adv. Neural Inf. Process. Syst.* **2012**, *25*, 1097–1105. [CrossRef]
2. Szegedy, C.; Ioffe, S.; Vanhoucke, V.; Alemi, A. Inception-v4, Inception-ResNet and the impact of residual connections on learning. In Proceedings of the Thirty-first AAAI Conference on Artificial intelligence, San Francisco, CA, USA, 4–9 February 2017; pp. 4278–4284.
3. Chaib, S.; Yao, H.; Gu, Y.; Amrani, M. Deep feature extraction and combination for remote sensing image classification based on pre-trained CNN models. In Proceedings of the Ninth International Conference on Digital Image Processing (ICDIP 2017), Hong Kong, China, 21 July 2017; p. 104203D.
4. Vandal, T.; Kodra, E.; Ganguly, S.; Michaelis, A.; Nemani, R.; Ganguly, A.R. Deepsd: Generating high resolution climate change projections through single image super–resolution. In Proceedings of the 23rd ACM SIGKDD International Conference on Knowledge Discovery and Data Mining, Halifax, NS, Canada, 9 August 2017; pp. 1663–1672.
5. Young, T.; Hazarika, D.; Poria, S.; Cambria, E. Recent trends in deep learning based natural language processing. *IEEE Comput. Intell. Mag.* **2018**, *13*, 55–75. [CrossRef]
6. Bengio, Y.; Ducharme, R.; Vincent, P.; Jauvin, C. A neural probabilistic language model. *J. Mach. Learn. Res.* **2003**, *3*, 1137–1155.
7. Devlin, J.; Chang, M.W.; Lee, M.K.; Toutanova, K. Bert: Pre-training of deep bidirectional transformers for language understanding. *arXiv* **2018**, arXiv:1810.04805.

8. Silver, D.; Huang, A.; Maddison, C.J.; Guez, A.; Sifre, L.; Driessche, G.V.D.; Schrittwieser, J.; Antonoglou, I.; Panneershelvam, V.; Lanctot, M.; et al. Mastering the game of Go with deep neural networks and tree search. *Nature* **2016**, *529*, 484–489. [CrossRef]
9. Vinyals, O.; Babuschkin, I.; Czarnecki, W.M.; Mathieu, M.; Dudzik, A.; Chung, J.; Choi, D.H.; Powell, R.; Ewalds, T.; Georgiev, P.; et al. Grandmaster level in StarCraft II using multi-agent reinforcement learning. *Nat. Cell Biol.* **2019**, *575*, 350–354. [CrossRef]
10. Moore, G. Cramming More Components onto Integrated Circuits (1965). In *Ideas That Created the Future*; The MIT Press: Cambridge, MA, USA, 2021; Volume 38, pp. 261–266. [CrossRef]
11. Gams, M.; Gu, I.Y.-H.; Härmä, A.; Muñoz, A.; Tam, V. Artificial intelligence and ambient intelligence. *J. Ambient. Intell. Smart Environ.* **2019**, *11*, 71–86. [CrossRef]
12. Yampolskiy, R.V. *Artificial Superintelligence: A Futuristic Approach*, 1st ed.; CRC Press: Boca Raton, FL, USA, 2015.
13. Gams, M.; Kolenik, T. Relations between Electronics, Artificial Intelligence and Information Society through Information Society Rules. *Electronics* **2021**, *10*, 514. [CrossRef]
14. Tervonen, J.; Pettersson, K.; Mäntyjärvi, J. Ultra-Short Window Length and Feature Importance Analysis for Cognitive Load Detection from Wearable Sensors. *Electronics* **2021**, *10*, 613. [CrossRef]
15. Kompara, T.; Perš, J.; Susič, D.; Gams, M. A One-Dimensional Non-Intrusive and Privacy-Preserving Identification System for Households. *Electronics* **2021**, *10*, 559. [CrossRef]
16. Brena, R.; Escudero, E.; Vargas-Rosales, C.; Galvan-Tejada, C.; Munoz, D. Device-Free Crowd Counting Using Multi-Link Wi-Fi CSI Descriptors in Doppler Spectrum. *Electronics* **2021**, *10*, 315. [CrossRef]
17. Lin, H.-C.K.; Ma, Y.-C.; Lee, M. Constructing Emotional Machines: A Case of a Smartphone-Based Emotion System. *Electronics* **2021**, *10*, 306. [CrossRef]
18. Bednarek, M.; Kicki, P.; Bednarek, J.; Walas, K. Gaining a Sense of Touch. Object Stiffness Estimation Using a Soft Gripper and Neural Networks. *Electronics* **2021**, *10*, 96. [CrossRef]
19. Bednarek, M.; Kicki, P.; Walas, K. On Robustness of Multi-Modal Fusion—Robotics Perspective. *Electronics* **2020**, *9*, 1152. [CrossRef]
20. Mańkowski, T.; Tomczyński, J.; Walas, K.; Belter, D. PUT-Hand—Hybrid Industrial and Biomimetic Gripper for Elastic Object Manipulation. *Electronics* **2020**, *9*, 1147. [CrossRef]

Review

Relations between Electronics, Artificial Intelligence and Information Society through Information Society Rules

Matjaž Gams [1] and Tine Kolenik [1,2,*]

1. Department of Intelligent Systems, Jožef Stefan Institute, 1000 Ljubljana, Slovenia; matjaz.gams@ijs.si
2. Jožef Stefan International Postgraduate School, 1000 Ljubljana, Slovenia
* Correspondence: tine.kolenik@ijs.si

Abstract: This paper presents relations between information society (IS), electronics and artificial intelligence (AI) mainly through twenty-four IS laws. The laws not only make up a novel collection, currently non-existing in the literature, but they also highlight the core boosting mechanism for the progress of what is called the information society and AI. The laws mainly describe the exponential growth in a particular field, be it the processing, storage or transmission capabilities of electronic devices. Other rules describe the relations to production prices and human interaction. Overall, the IS laws illustrate the most recent and most vibrant part of human history based on the unprecedented growth of device capabilities spurred by human innovation and ingenuity. Although there are signs of stalling, at the same time there are still many ways to prolong the fascinating progress of electronics that stimulates the field of artificial intelligence. There are constant leaps in new areas, such as the perception of real-world signals, where AI is already occasionally exceeding human capabilities and will do so even more in the future. In some areas where AI is presumed to be incapable of performing even at a modest level, such as the production of art or programming software, AI is making progress that can sometimes reflect true human skills. Maybe it is time for AI to boost the progress of electronics in return.

Keywords: information society; electronics; artificial intelligence; ambient intelligence

Citation: Gams, M.; Kolenik, T. Relations between Electronics, Artificial Intelligence and Information Society through Information Society Rules. *Electronics* **2021**, *10*, 514. https://doi.org/10.3390/electronics10040514

Academic Editor: Nikolay Hinov

Received: 5 January 2021
Accepted: 18 February 2021
Published: 22 February 2021

Publisher's Note: MDPI stays neutral with regard to jurisdictional claims in published maps and institutional affiliations.

Copyright: © 2021 by the authors. Licensee MDPI, Basel, Switzerland. This article is an open access article distributed under the terms and conditions of the Creative Commons Attribution (CC BY) license (https://creativecommons.org/licenses/by/4.0/).

1. Introduction

What are the relations between information society (IS), electronics and artificial intelligence (AI)? In this paper we first introduce description of AI and related fields, and then proceed to the information society as the most general concept.

The term "artificial intelligence", also called "machine intelligence" (MI), includes hardware, software or most common combined artificial systems, i.e., machines that exhibit some form of intelligence. For example, a mobile phone runs a game or provides a web search using algorithms written in a programming language that is actually running on a mobile phone or in a cloud.

"Ambient intelligence" (AmI) is AI implemented on machines in the surrounding environment and is mostly demonstrated by the services of the human environment. AmI is therefore even closely related to various machines, not necessarily computers, taking care of humans thus benefiting from intelligent and cognitive functionalities.

Both AI and AmI are characterized as the study of intelligent agents [1], representing a core building block of all AI and AmI methods. An intelligent agent is a system that perceives its environment and takes actions. There are several levels of agents ranging from the simplest reflective ones such as a thermostat to advanced ones that learn and follow their goals, e.g., autonomous cars [2].

The term AI was first coined by McCarthy in 1956 at the Dartmouth conference. In Europe, several researchers consider 1950 to be the start of AI, which was then termed machine intelligence. The milestone was Alan Turing's paper [3] published in 1950. Alan

Turing is regarded by many as the founding father of computer science due to his halting problem, the Turing test, the Turing machine and the decoding of Hitler's Enigma machine that helped to end the Second World War more quickly [3].

The term AmI was introduced in the late 1990s by Eli Zelkha and his team at Palo Alto Ventures [4] and was later extended to the environment without people. A modern definition was delivered by Juan Carlos Augusto and McCullagh [5]: "Ambient Intelligence is a multi-disciplinary approach which aims to enhance the way environments and people interact with each other. The ultimate goal of the area is to make the places we live and work in more beneficial to us." AmI is also aligned with the concept of the "disappearing computer" [6,7]. The AmI field is very closely related to pervasive computing, ubiquitous computing and context awareness [8–11].

AI and AmI are two of the most prosperous fields in the current area of human civilization, named information society (IS). In its most general form, an "information society" can be characterized as a society in which any kind of activity regarding information is an integral and inseparable part. There are many activities that involve information as the primary object: use, construction, manipulation, processing, integration, recording, accessing, storage, transfer, etc. An information society was first defined as a society in in which 30% of its gross domestic product (GDP) relied on information, but the definition was already met for most of the developed countries decades ago [12]. Historically, information had been regarded as a valuable source of any kind of progress, and information became the essential driver of an IS by being able to be dispersed and processed exponentially faster and in larger quantities than ever before with communication technologies. The rapid, radical and thorough change that these capabilities offered to society changed societies from industrial to information societies. The emergence of the importance of information has reorganized education, the economy, healthcare, warfare, governmental services and democratic operations, industry, scientific investigation and other, lower-level aspects of what is deemed important in a society. As a prominent example, medicine has seen exponential progress in screening and diagnosing of a wide range of diseases, which can mostly be attributed to AI and its use in medical imaging [13–15]. This fundamentally changed the populace as the main participants in the change, who entered this new phase of a civilization through accessing the Internet [16].

The Internet was therefore one of the main drivers of the steady change that followed its adoption. Most theoreticians pinpoint the start of the IS as the 1970s, which was when the Internet started to be used internally [17]. However, it was the early 1990s and 2000s that brought the most changes and rapid progress through net usage and information dissemination. More widely, information and communication technologies (ICTs) and their usage intensified how we centre our activities in economic, social, cultural and political areas [18]. While the start of the IS can be pinpointed as the early 1970s, the most common recent or golden IS era due to the growth in the amount of data started in approximately 2000 (see Figure 1). The promise of this change was reaffirmed by an international document signed in 2005 called the Tunis Agenda for the Information Society [19], which called for strong financing of ICTs. The reason for such a strong systemic action by many nations was that the use of ICTs resulted in progress for social good and in progress that benefitted the populace; therefore, is was seen as the new foundational grounds for progressing towards a new age in which the IS would be the standard. This progress results in the fast growth of Internet users, as presented in Figure 2. As a consequence, employment needs are shifting and influencing human everyday life.

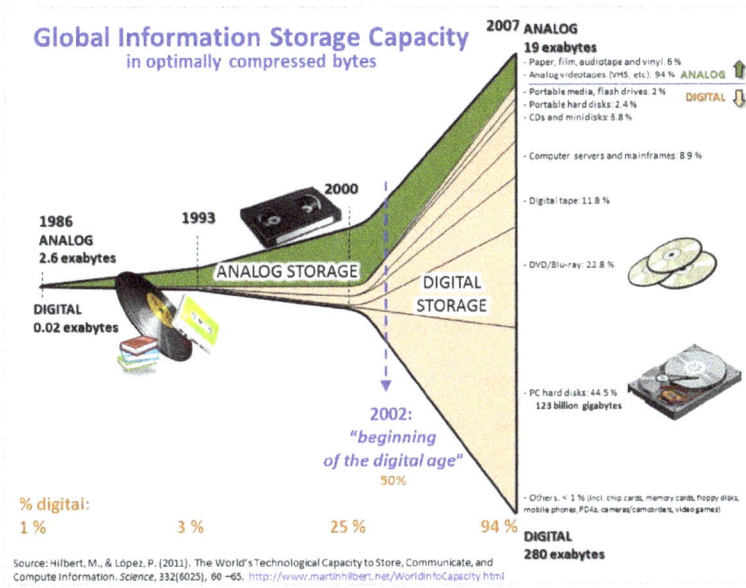

Figure 1. The exponential growth of data in the information society [20].

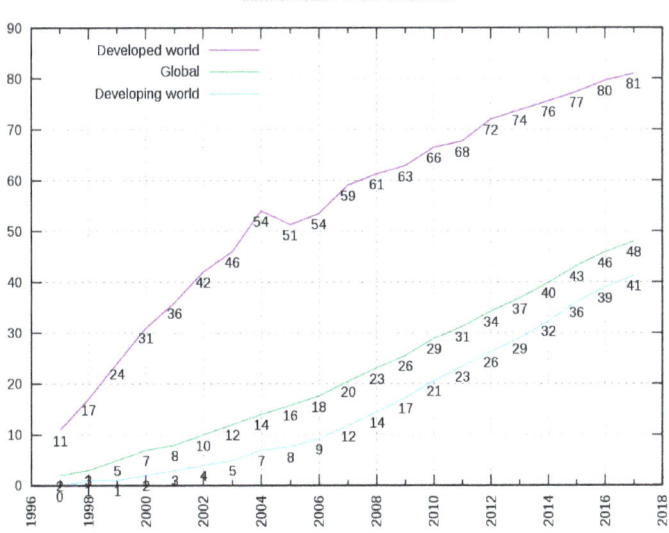

Figure 2. Growth in the number of Internet users [21].

In this paper, we concentrate on the IS laws related to electronics, information, IS, AI and ambient intelligence (AmI) as a sister field of AI. Section 2 briefly describes the related work, Sections 3 and 4 concentrate on IS laws, Section 5 continues by describing the longevity of Moore's law, Section 6 describes the relations to artificial intelligence [22] and ambient intelligence [23], and Section 7 ends the paper with the conclusion and limitations of the presented analysis.

2. Related Work

In this section, the related work pertaining to IS laws and their connection to AI is presented. To the best of our knowledge, the work this paper overviews has not been presented in this way before. However, there are some overview papers that are related to the topic at hand that should be mentioned. Some only observe the laws but do not name, codify and define them as this paper does. Webster [18] wrote an academic book on the theories of an information society, and it addresses some areas of the laws described in this paper: namely, how an IS relates to democracy (in this paper, Clift's law), how technology propelled us from an industrial to an information society (whose process is described by the laws in the paper), and how the abundance of information influences people (e.g., infobesity, Gross' law). Boyle, in his academic book Shamans, Software & Spleens [24], discusses the influence of an IS on economics (see Section 4.2) and liberalism (Tapscott's law in this paper). Connecting an information society to growth and possibilities for artificial intelligence has also been analysed. The area that received the most attention is privacy in the era of artificial intelligence and how easy it is to obtain personal data and to exploit them [25,26]; furthermore, some works emphasize the growth of artificial intelligence in IS [24,27,28].

The special issue of the *Computer* journal in 2013 [29] presented several basic electronic laws analysed in this paper. While seven years old, several laws and issues presented are still valid now. Furthermore, some issues in the *Computer* journal are studied to a greater and more specific level than the condensed descriptions in this paper. However, the special issue did not present an explicit list or electronic rules, and the number of laws was significantly lower.

The initial work leading to this paper was first presented in 2002 in [30] and was systematically lectured at the university level, progressing each year.

Finally, AI and AmI are analysed in [31] and resemble some relation to Section 5 in this paper.

Table 1 summarizes the described related work and the IS topic they tackle.

Table 1. Summarized related work and their characteristics, relevant for this paper.

Related Work	Relevant Characteristics
[18]	IS related to democracy, Clift's law, Gross' Law
[24]	IS related to economics and liberalism, Tapscott's law
[25,26]	Privacy in the era of AI and its exploitation
[24,27,28]	General growth of AI
[29]	First presentation of a smaller number of IS laws in an academic form (not a list or in terms of IS progress)
[30]	IS technology that enables current society
[31]	AI and AmI analysis

As evidenced, the related work addresses more specific areas of IS, AI and AmI, without their firmer relation to electronics, while this paper encompasses the laws from a specific perspective, some first presented in such a form, from many distinct areas of an IS in a novel comprehensive list and relates the list to trends of AI, AmI, an IS and human progress in general, analysing their impact and predicting future directions.

3. Basic Information Society Laws

The progress of an information society can be presented through several IS laws or computing laws [29,30,32]. In this paper, however, the emphasis is on the IS laws as descriptors of the relations to IS, AI and electronics. The laws are therefore grouped into basic technological laws presented in this section, and software laws, socioeconomic laws, and computer and progress laws in the next section. Several of the laws are related to AI in this or another way:

1. Moore's law [33]: The growth of the capabilities of electronic devices, e.g., chips, is exponential. As originally stated, the number of transistors in a dense integrated circuit (IT) doubles approximately every two years. This law has been valid for over half a century and is analysed in more detail later.
2. Joy's law [34]: The peak computer speed doubles each year. This law was first formulated in 1983. The formula is ComputerSpeed = 2 ** (year: 1984). The rule is related to Moore's law and bears the same time-resistant properties.
3. Pollack's law [35]: Due to microarchitectural advances, microprocessor performance increases roughly proportional to the square root of the increase in complexity, whereas power consumption increases roughly linearly proportional to the increase in complexity. Pollack's law implies that microarchitectural advances improve the performance by $\sqrt{2} \approx 41\%$, thus bearing some similarity to Moore's law and allowing progress without exceeding the energy demands.
4. Bell's law [36]: Roughly every decade, a new, lower priced computer class (or generation) forms based on a new programming platform, network and interface, resulting in new usage and the establishment of a new industry. It is related to the Moore's law, which refers to years; however, Bell's law refers to computer classes, i.e., generations. It takes approximately 10 years to exploit the possibilities of a particular computer class, and during that time a new computer class is researched and finally introduced.
5. Kryder's law [37]: Disk capacity (more specifically, magnetic disk areal storage density) grows exponentially, even faster than Moore's law. However, similar to Moore's law but much sooner, the limit of fast growth was achieved in approximately 2019, and the magnetic disk capacity was then more or less stable [38].
6. Makimoto's law [39]: There is a 10-year cycle between research and standardization, meaning that we can see future commercial capabilities by examining today's research facilities. There is also Makimoto's wave [40], which explains not only the semiconductor waves but also the AI and machine learning (ML) waves. Indeed, AI has progressed in waves, but not exactly in 10-year waves. Unlike Moore's law, Makimoto's law describes the general property between research and the market in electronics and is not as prone to time as some other rules.
7. Keck's law [41]: Communication capabilities (actual traffic) grow exponentially. Keck's law has successfully predicted the trends for the data rates in optical fibres for four decades. Keck's law is another example of an exponential law predicting incredibly fast growth that was valid for a certain time period but is currently slowing and may be facing a plateau in the foreseeable future.
8. Gilder's law or the law of telecoms [42]: The total telecommunications system capacity (b/s) triples every three years, and the bandwidth grows at least three times faster than computing power. Gilder's law is similar to Keck's law.
9. Koomey's law [43]: The number of computations per joule of energy dissipated has been doubling approximately every 1.57 years. Similar to other exponential laws, Koomey's law is losing its consistency. In 2000, the doubling slowed to every 2.6 years. Koomey's law is also related to the end of Dennard scaling in 2005, i.e., the ability to build smaller transistors with constant power density.
10. Dennard's law or Dennard scaling [44]: As the size of transistors decrease, their power density stays constant. It is strongly related to one period of Moore's law but is more or less saturated.
11. Rock's law or Moore's second law [45]: The cost of a semiconductor chip fabrication plant doubles every four years. This law is related to technological progress, although without the past issue with time validity as Moore's (first) law.
12. Neven's law [46]: Quantum computers are gaining computational power at a doubly exponential rate. Quantum supremacy was declared by Google in October 2019. In October 2020, quantum supremacy was reclaimed by Chinese researchers [47], but both publications raised several questions. The law claims that quantum computers

are progressing fast, thus enabling further growth of computational computing power. The timescale of this rule has yet to be observed for a sufficient number of years.

13. Amdahl's law [48]: Amdahl's law predicts the theoretical speedup limit when using multiple processors, meaning there is always a fraction of a problem that cannot be parallelized. It can be defined with the following formula defining speedup S using the percentage p of the tasks that can be parallelized and the availability of threads s that enable parallel execution: $S = 1/(1 - p) + p/s)$. At the limit, when there is an unlimited supply of parallel execution mechanisms, this equation turns into $1/(1 - p)$. Amdahl's law is not sensitive to time-related issues.
14. Gustafson's or Gustafson–Barsis's law [49]: This law addresses the shortcomings of Amdahl's law by considering flexible tasks and is more accurate for faster devices.
15. Grosch's or Cray's law [50]: Computing performance or added economy corresponds to the square root of the increase in speed; that is, to perform a calculation 10 times as cheaply, you must perform it 100 times as fast. The law is not directly related to advances in microelectronics and might be time-independent, but more future data are needed to confirm it.

4. Related Information Society Laws

The IS laws presented in this section deal with relations to human activities, in particular in regards of the usage of computers, human–computer or human–human interactions.

4.1. Software Laws

1. Linus's law [51]: Given large enough beta tester and codeveloper bases, almost every problem will be characterized quickly and the fix will be obvious to someone. In other words: "given enough eyeballs, all bugs are shallow". The law contradicts fears that software is becoming uncontrollable with the growing amount of code and is not related to technological issues but to human ingenuity. As such, the law seems to be quite time-independent.
2. Wirth's, Page's, Gates' or May's law [52]: Software is becoming slower more rapidly than hardware is becoming faster. This law may not be fully confirmed in recent years due to various tools and new techniques, and in particular the time relations in this rule seem to be under consideration.
3. Brooks's law: "Adding manpower to a late software project makes it later" is an observation about software project management, but is valid in several other areas where the process cannot be parallelized. It was coined by Fred Brooks in his 1975 book The Mythical Man-Month [53]. Somewhat ironically, an incremental person, when added to a project, makes it take more, not less time.

4.2. Socioeconomic Laws

1. Metcalf's law [54]: Value of a network = square(n), where n is number of nodes in the network; the value or effect of a network is proportional to the square of the number of nodes. It is similar to Odlyzko's law [55]: value of a network = n × logn. This law seems to be time-independent.
2. Tapscott's or Negroponte's law: The economy of an information society is a frictionless economy, information economy, Internet economy, net economy, and new economy; it is global, liberal, without restrictions and regulations, spurred by electronics and information technologies and based on bits instead of atoms. Tapscott [56] introduced e-commerce and e-business characteristics, while Negroponte [57] introduced the e-world consisting of bits instead of atoms, transforming the economy into a new stage. The economy in an IS is surely different compared to the period before, but it will last only a certain period of time until another step in human civilization occurs [58].
3. Gross's law: The information overload law; or the infobesity, infoxication, information anxiety and information explosion law [59]: The side effect of an IS is information

overload. This law relates to the excessive information given to people in everyday life and when making decisions due to ICTs generating massive amounts of information that grow exponentially. This law seems to be increasingly more valid with the progress of increasingly more data and information and with the lack of appropriate mechanisms that would enable people to handle the information overload issue.

4. Gams' law [30]: IS, the cyberworld double fortune. The fortune can be real or fictitious, such as cryptocurrency. First presented in 2002, when there was not as much cryptocurrency in the world such as Bitcoin, the observed law taught among the local economics faculty proposes a transition at a remote island where native people trade natural goods such as pigs and coconuts. At one point, a modern king introduces paper money; in their fictitious currency, 1 Illa is worth 1 pig. Counting the natural resources and the paper money, the island has twice as much wealth as before. If neighbouring islands accept their currency, the king can print considerably more paper money and buy a substantial amount of goods abroad. In time, the king's successor introduces BIlla, a Bitcoin version of their paper currency Illa. The story repeats and the current king, or better, their business elite, can considerably increase their worth. This example should help understand the events in the net economy: why virtual money increases wealth, why elites become increasingly richer and why the fictitious or "normative" standard may not directly correspond to the real status of netizens. For example, the netizens on the fictitious island have the same amount of pigs and coconuts at the end of the story as in the beginning, and if the elites increase their wealth, the average islander has less than in the beginning. Note, however, that the progress enables better production of pigs and other goods, and overall, the middle class more or less stays at the same level while the overall wealth increases. However, nominal wealth is significantly different than actual wealth in terms of pigs and coconuts. As with many economic laws, this one is also not directly bound to the technological process and therefore is not as time-dependent as, e.g., Moore's law.

5. Clift's law or e-democracy, digital democracy or Internet democracy progress [60]: The web enables democratic progress. The introduction of ICTs and IS tools to political and governance processes is thought to promote democracy since citizens are presumed to be eligible to participate equally in information creation and sharing. In other words, "The Internet is the most democratic and free media in the world." The World Wide Web supposedly offers participants "a potential voice, a platform, and access to the means of production" [61]. However, in recent decades, the concentration of capital has resulted in an increased concentration of media ownership by large private entities in several American and European countries [62]. According to current polls [63], over 90% of Americans from a sample of approximately 20,000 considered the media to have major importance for democracy; however, approximately 50% of them see the media as biased to various degrees, impairing and endangering democratic processes. While for decades the optimistic viewpoint prevailed in e-democracy, in recent years, we might be witnessing a change. The future of this law seems quite unclear.

4.3. Computers and Progress Laws

When researchers from MIT tested 62 technologies [64], they found that there are two laws that fit the data very well, Moore's law and Wright's law, although one is related to computers and the other is related to flight. Consider Wright's law:

1. Wright's law [65]: The cost of airplanes is proportional to the inverse of the number of planes manufactured raised to some power. The law seems to be time-independent.

This theory was proposed as a more general law that governs the costs of technological products and is often explained on the basis that the more of a product that we make, the better and more efficient we become at making the product. It has also been shown [66] that costs decrease because of economies of scale.

The analysis in [64] further compared the two most consistent laws, Wright's and Moore's law. It was stated that if the production of an item grows at an exponential rate, then Wright's law and Moore's law are quite similar. This is slightly hard to comprehend since computing is often regarded as a special case ("It's a much more general thing," says author Doyne Farmer, currently at the University of Oxford, United Kingdom) and that Moore published the law from purely a technological observation based on the progress of electronics. Indeed, the consequences of fast exponential growth such as that observed by Moore in computing are in fact related to the technology life cycle, which describes the commercial gain of a product through research and development expenses and the financial return during its profitable stage [67].

The commercial effects of Moore's law in IS products are indeed fascinating. For example, the total production of semiconductor devices resulted in one transistor produced per metre of our galaxy's diameter and billions of transistors per star in our galaxy. The calculation is as follows: according to [29], transistor production reached 2.5×10^{20}, which is 250 billion billion, in 2014 [68]. Our Milky Way galaxy has a diameter between 100,000 and 180,000 light years [69]. The galaxy is estimated to contain 100–400 billion stars and 100 billion planets [70]. In addition, there are also approximately 100 billion neurons in our brain [71].

5. Longevity of Moore's Law

Moore's first original law is focused on the transistors in an IC. The law states that the number of transistors in such a circuit doubles every two years with the circuit remaining the same size [72]. The original Moore's law and the consequent ones are based on historical observations and assume that the property will last for a reasonable time period. Although it is called a law, it is not a law in the sense of physics laws, e.g., $F = m \times a$. Moore's first law was revised and applied repeatedly to different areas: microprocessors, memory capacity, sensors and pixilation [72] The revisions of the law were proposed (i) to encapsulate additional properties of technology, which resulted in performance changes, manufacturing processing and costs and other general market trends; and (ii) to avoid problems with specific saturations of technological parameters while at the same time continuing the exponential progress of computing power.

The law proved to be very consistent in its dynamic form for a long time, enabling reliable predictions. In semiconductor production and other listed areas where Moore's law was applied, the law helped guide planning, setting targets, scheduling and other processes related to organization, research and development [72].

While there are some limitations to any process, they must be carefully evaluated. For example, consider the share of Internet users in the population. Figure 1 shows fast growth and that the upper limit is 100%, meaning the end of progress. However, even in the case in which all humans use the Internet, there will be a growing number of devices and intelligent systems using the Internet, thus increasing the absolute number of overall users. Such misconceptions are common and demand attention when various sources of information are encountered. Similarly, warnings of computer progress stalling were issued years and decades ago [16]. Even the first author of the paper was warned 40 years ago by a specialized professor in the computer science faculty that the pace is slowing, and the warning seemed to be validated by the data and forthcoming saturation. Some of the saturations can be observed in Figure 3 [72]. However, the figures on progress in the 55-year-long life of Moore's law show no overall downgrade of the exponential volatile growth. For example, microprocessor architects report that semiconductor advancement has slowed below the pace predicted by Moore's law since 2010. However, as of 2018, leading semiconductor manufacturers have developed mass production processes for IC fabrication that are claimed to keep pace with Moore's law. Presently, there is not a consensus among experts on exactly when Moore's law will cease to apply.

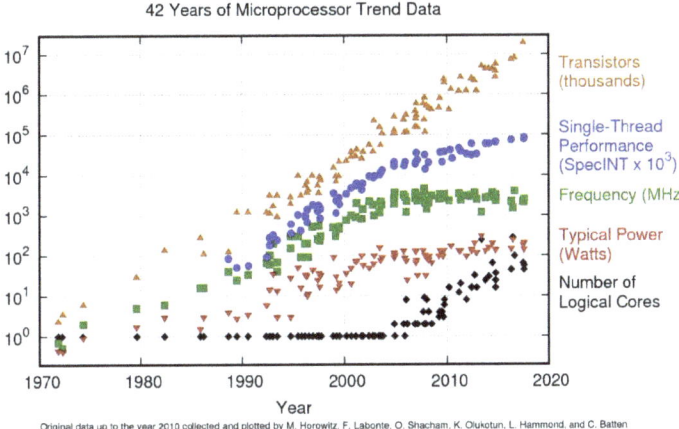

Figure 3. Some saturations related to Moore's law [72].

In reality, Moore's law is rather stable in terms of the annual computing gain in chip capacities of 50%, even though several partial Moore's laws have already ended. What happened is that one approach to increase cheap performance was followed by another, successfully continuing the overall Moore's law, e.g., designing larger chips with more layers when the technology of one chip hit a wall. Currently, there are several possibilities to continue Moore's law for several decades, such as new technologies including quantum computing or 3D chips and sophisticated algorithms.

One of the three most advanced chip companies, Intel, has a team of 8000 hardware engineers and chip designers whose jobs and careers depend on chip progress. While the end of growth was predicted decades ago, they were able to find ample technical opportunities for advances. They estimate that there are probably more than a hundred variables involved in keeping Moore's law going; their director says he has been hearing about the end of Moore's law for his entire career. After a while, he "decided not to worry about it" [5].

There is a limit to any progress, and the introduction of a new device or a new type of device generally follows an S-curve (see Figure 4): after a slow start, it grows exponentially and finally becomes saturated.

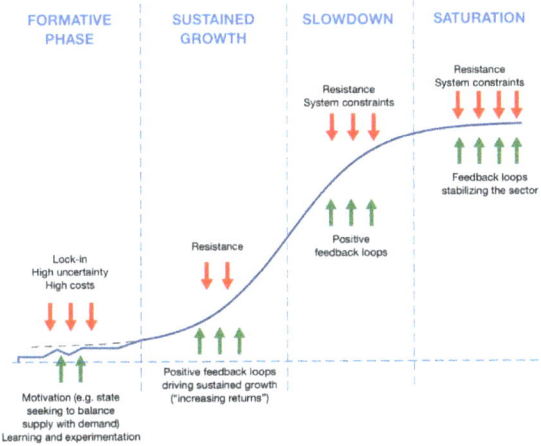

Figure 4. S-curve for the introduction of a new product [73].

Regardless of the future, history has demonstrated that the progress of electronics so far is beyond saturation. Every year since 2001, the MIT Technology Review yearly has provided the ten most important breakthrough technologies. Practically all proclaimed technologies are possible only because of the computation advances described by Moore's law, and they also fuelled breakthroughs in artificial intelligence, communication and genetic medicine by giving machine-learning techniques the ability to process massive amounts of data to find answers [74].

In summary, what we can claim with high certainty is that information communication technologies significantly and probably spurred human progress in the last century the most out of all technologies [58] and that the progress will continue at a similar pace in the decade or two to come. In addition, this progress has and will largely influence the progress of artificial intelligence and ambient intelligence.

6. AI, AmI and Electronics

In this section, AI and its relation to IS and IS laws is presented. Understanding the intertwined progress of AI is essential since it already influences practically every human activity and will do so even more in near future.

The history of AI and AmI is characterized by cycles of overoptimism and overpessimism, often referred to as the "AI winter" [75] (Figure 5), somewhat resembling the stalling of a particular technology enabling Moore's law. There are a couple of differences, however. Whereas there were papers stating that Moore's law is coming to an end several times before, the funding in electronics never dried out. Instead, the AI funding varied a lot in each case when a particular AI approach such as expert systems met its limitations. Unlike the progress of electronics where the pessimism was local and not wide spread in public, the AI winter was generally accepted in our society as a terminal inability to perform similar to humans. In computing, there is an old saying: "Never say never", since so many negative predictions failed. Therefore, the AI pessimism was ill-founded from the start since it was only a question of time when advanced electronics would promote AI growth. Whatever the case, the development of new chips and new AI methods soon enabled fast and prosperous progress of information society.

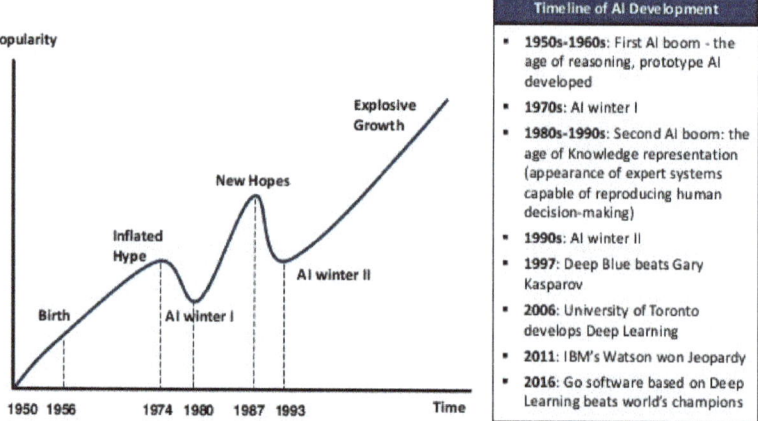

Figure 5. AI periods of overoptimism and overpessimism [76].

The ability of the broad AI field to solve more difficult problems in time is basically guaranteed by the exponential growth of computing semiconductor capabilities, as defined by IS laws. At the same time, AI progress is also characterized by methodological breakthroughs in bursts. Currently, AI and AmI achievements attract worldwide attention in practically all areas of academia, gaming, industry and real life.

AI and AmI have already penetrated every aspect and are already having large impacts on our everyday lives. Smartphones, all modern cars, autonomous vehicles, the Internet of Things (IoT), smart homes and smart cities, medicine and institutions such as banks or markets all use artificial intelligence on a daily basis [77]. Examples include when we use Siri, Google or Alexa to request directions to the nearest petrol station or to order a taxi, when we make a purchase using a credit card with an AI system in the background checking for potential fraud, when an intelligent agent in a smart home regulates user-specific comfort, or when a smart city optimizes heterogeneous sources and environment demands.

When fed with huge numbers of examples and with fine-tuned parameters, AI methods and, in particular, deep neural networks (DNNs) regularly beat the best human experts in increasing numbers of artificial and real-life tasks [78], such as diagnosing tissue resulting from several diseases. There are other everyday tasks, e.g., the recognition of faces from a picture, where DNNs recognize hundreds of faces in seconds, a result no human can match. Figure 6 demonstrates the progress of DNNs in visual tasks. In approximately 2015, visual recognition using neural networks was slightly better than that of humans in specific domains; now, neural networks have surpassed humans quite significantly in several visual tests, for example, fast recognition of faces in foggy conditions. Research on how human vision works using AI that can inform future advances in artificial vision has also made strides [79]. In one of the core areas of AmI research, human-activity recognition, the application of deep-learning algorithms also achieved human-level results in recent years (see [10,80–82]).

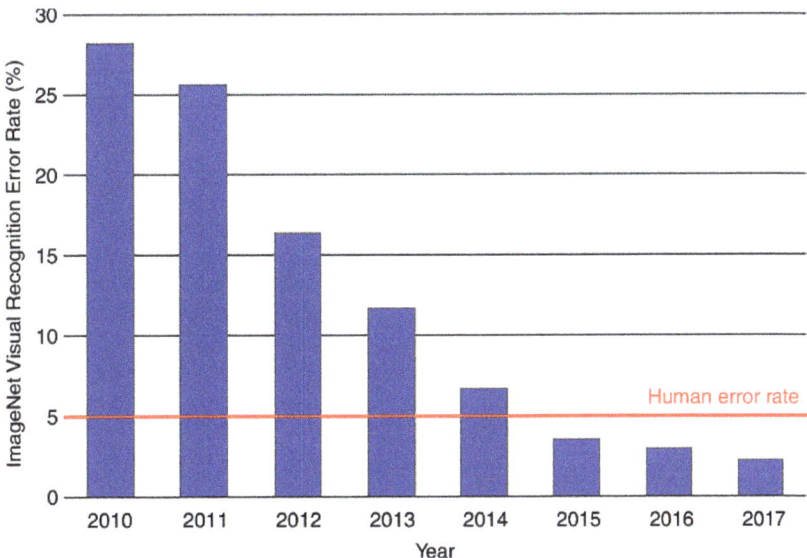

Figure 6. Error rates of deep neural networks (DNNs) on the ImageNet competition over time [83].

DNNs make it possible to solve several practical tasks in real life, such as detecting cancer [84–86] or Alzheimer's disease [87–89]. Furthermore, studies applying DNNs to assess facial properties can reveal several diseases, sexual orientation, intelligence quotient (IQ) and political preferences. One of the essential tasks of AmI is to detect the physical, mental, emotional and other states of a user [90–92]. In real life, it is important to provide users with the comfort required at a particular moment without demanding tedious instructions.

When will AI systems outperform humans in nearly all properties? The phenomenon is termed superintelligence (SI) or sometimes superartificial intelligence (SAI), and it is often related to singularity theory [93]. The seminal work on superintelligence is Bostrom's Superintelligence: Paths, Dangers, Strategies [94]. Another interesting, more technically oriented analysis is presented in the book Artificial Superintelligence: A Futuristic Approach by R. V. Yampolskiy [11].

Similar to superintelligence, there is also superambient intelligence (SAmI). SAmI is more strongly related to humans and therefore possibly more likely to approach human-related intelligence. For example, when taking care of humans and handling the environment, a user must feel as mentally and cognitively comfortable as possible; thus, AmI must understand human cognitive [95], emotional and physical current states and not just environmentally related properties such as energy consumption or safety. In addition, while hostile superintelligence in movies often evolves over the Web, in reality, SAmI would even more easily transverse in the environment of smart devices, homes and cities. However, there seems to be no motivation for hostility between an IS and humans as long as we avoid risky activities such as autonomous weapons [96].

Finally, now might be the time for AI to pay back the merit for fast progress; for decades, electronics dominated by Moore's law stimulated fast AI progress, and now AI can help build better chips, computers and other electronic devices. There are several possibilities, e.g., the internal architecture of chips is rather simple and can be tuned to specific tasks such as vision or programming. Even programming itself can already be speeded by a factor of 1000 by smart transformation of a program in Python into an optimized C version. While it seems a distant future to code such complex tasks for current systems such as GPT-3 (Generative Pre-trained Transformer 3), the progress in recent years is more than promising [97].

7. Conclusions

This paper introduces relations between electronics, the information society and artificial intelligence as three highly interdependent and intertwined fields. The comprehensive list of IS laws on its own represent a novel contribution since the laws described in various encyclopedias do not provide an overview on these fundamental relations. There are also additional analyses of their progress and longevity. Most likely, the only well-known law among the public is Moore's law, but even there there are still some misinterpretations in the general media, e.g., the overpessimistic viewpoints that claim that the law will soon end, as was proclaimed several times before when a particular single mechanism enabling computer capability growth reached the stage of saturation. The overoptimistic viewpoints, on the other hand, expect unlimited growth for an indefinite time period, which seems practically and theoretically impossible since sooner or later any progress reaches its limits, especially exponential growth, which typically follows S-curve dynamics (see Figure 4).

Many IS laws share the fate of Moore's law: they describe exponential growth and will sooner or later, if they have not already, reach saturation. Regardless of the case, exponential growth influences the growth of AI and AmI and the progress of human civilization. The potential stalling of the progress of electronics is not good news, but luckily there are hundreds of ways to continue the current exponential improvement of the capabilities of electronic computing components.

The list of IS laws also shows the highly interconnected relations to AI and progress of our civilization, which is now in the stage of information society. Among other technologies, AI and AmI are already enabling various improvements not possible without them, including the design of new computing devices. While electronics enabled the progress of fields such as computing or AI, the tables are turning. AI and human ingenuity combined are now to enable future progress of electronics for at least decades to come, thus enabling further progress of human civilization.

Our future work will try to address the limitations of this paper. Since some IS laws are only now reaching their saturation point, further observations and analyses are needed to predict how they will behave in the near future, and which laws will emerge to replace them. Such detailed analyses are out of scope for this paper, as analysing just one such law necessitates its own paper. Due to the dynamic nature of the AI evolution, as seen with AI winters and then sudden bursts of AI advancement, it is hard to pinpoint the relations between the forthcoming IS laws in AI, as they do not affect each other linearly. The future might bring us anything from faster progress to saturation or even decline in case of catastrophic events, such as a significantly deadlier pandemic than COVID-19. This paper should therefore serve as a cardinal starting point with a presumption that the future will follow one of the predicted options.

Author Contributions: Conceptualization, M.G.; methodology, M.G. and T.K.; formal analysis, M.G.; investigation, M.G. and T.K.; resources, M.G.; IS laws, M.G.; writing—original draft preparation, M.G. and T.K.; writing—review and editing, M.G. and T.K.; visualization, T.K.; supervision, M.G.; project administration, T.K. All authors have read and agreed to the published version of the manuscript.

Funding: This research received no external funding.

Acknowledgments: The authors acknowledge the financial support from the Slovenian Research Agency (research core funding No. P2-0209).

Conflicts of Interest: The authors declare no conflict of interest.

References

1. Weiss, G. *Multiagent Systems (Intelligent Robotics and Autonomous Agents Series)*; MIT: Cambridge, MA, USA, 2013; ISBN 978-0262018890.
2. Russel, S.; Norvig, P. *Artificial Intelligence: A Modern Approach*, 3rd ed.; Pearson Education Limited: London, UK, 2014.
3. Turing, A. Computing Machinery and Intelligence. *Mind* **1950**, *59*, 433–460. [CrossRef]
4. Arribas-Ayllon, M. *Ambient Intelligence: An Innovation Narrative*; Lancaster University: Lancaster, UK, 2003. Available online: http://www.academia.edu/1080720/Ambient_Intelligence_an_innovation_narrative (accessed on 4 February 2021).
5. Augusto, J.C.; McCullagh, P. Ambient intelligence: Concepts and applications. *Comput. Sci. Inf. Syst.* **2007**, *4*, 1–27. [CrossRef]
6. Weiser, M. The computer for the twenty-first century. *Sci. Am.* **1991**, *165*, 94–104. [CrossRef]
7. Streitz, N.; Nixon, P. Special issue on 'the disappearing computer'. *Commun. ACM* **2005**, *48*, 32–35. [CrossRef]
8. Daoutis, M.; Coradeshi, S.; Loutfi, A. Grounding commonsense knowledge in intelligent systems. *J. Ambient. Intell. Smart Environ.* **2009**, *1*, 311–321. [CrossRef]
9. Ramos, C.; Augusto, J.C.; Shapiro, D. Ambient intelligence—The next step for artificial intelligence. *IEEE Intell. Syst.* **2008**, *23*, 15–18. [CrossRef]
10. Nakashima, H.; Aghajan, H.; Augusto, J.C. *Handbook of Ambient Intelligence and Smart Environments*; Springer: New York, NY, USA, 2009.
11. Yampolskiy, R.V. *Artificial Superintelligence: A Futuristic Approach*, 1st ed.; CRC Press: Boca Raton, FL, USA, 2015.
12. Soll, J. *The Information Master: Jean-Baptiste Colbert's Secret State Intelligence System*; University of Michigan Press: Michigan, WI, USA, 2014.
13. Artificial Intelligence in Medicine: Applications, Implications, and Limitations. Available online: http://sitn.hms.harvard.edu/flash/2019/artificial-intelligence-in-medicine-applications-implications-and-limitations/ (accessed on 4 February 2021).
14. Goyal, H.; Mann, R.; Gandhi, Z.; Perisetti, A.; Ali, A.; Aman Ali, K.; Sharma, N.; Saligram, S.; Tharian, B.; Inamdar, S. Scope of Artificial Intelligence in Screening and Diagnosis of Colorectal Cancer. *J. Clin. Med.* **2020**, *9*, 3313. [CrossRef] [PubMed]
15. Oren, O.; Gersh, B.J.; Bhatt, D.L. Artificial intelligence in medical imaging: Switching from radiographic pathological data to clinically meaningful endpoints. *Lancet Digit. Health* **2020**, *2*, e486–e488. [CrossRef]
16. Beniger, J.R. *The Control Revolution: Technological and Economic Origins of the Information Society*; Harvard University Press: Cambridge, MA, USA, 1986.
17. Byung-Keun, K. *Internationalising the Internet the Co-Evolution of Influence and Technology*; Edward Elgar Publishing: Cheltenham, UK, 2005.
18. Webster, F.V. *Theories of the Information Society*, 3rd ed.; Routledge: New York, NY, USA, 2006.
19. WSIS: Tunis Agenda for the Information Society. Available online: http://www.itu.int/net/wsis/docs2/tunis/off/6rev1.html (accessed on 10 December 2020).
20. Scholz, R. Sustainable Digital Environments: What Major Challenges Is Humankind Facing? *Sustainability* **2016**, *8*, 726. [CrossRef]
21. Global Internet Usage. Available online: https://en.wikipedia.org/wiki/Global_Internet_usage (accessed on 10 December 2020).
22. Finlay, S. *Artificial Intelligence for Everyone*; Relativistic: UK, 2020. ISBN 978-1-9993253-1-2.
23. Blokdyk, G. *Ambient Intelligence: A Complete Guide*; 5STARCooks: Huston, TX, USA, 2020.
24. Boyle, J. *Shamans, Software, and Spleens: Law and the Construction of the Information Society*; Harvard University Press: Cambridge, MA, USA, 1997.

25. Xiuquan, L.; Tao, Z. An exploration on artificial intelligence application: From security, privacy and ethic perspective. In Proceedings of the 2017 IEEE 2nd International Conference on Cloud Computing and Big Data Analysis (ICCCBDA), Chengdu, China, 28–30 April 2017; pp. 416–420. [CrossRef]
26. Lee, H.; Wong, S.F.; Oh, J.; Chang, Y. Information privacy concerns and demographic characteristics: Data from a Korean media panel survey. *Gov. Inf. Q.* **2019**, *36*, 294–303. [CrossRef]
27. Mansell, R.; Steinmueller, W. *Mobilizing the Information Society: Strategies for Growth and Opportunity*; Oxford University Press: Oxford, UK, 2020.
28. Cath, C.; Wachter, S.; Mittelstadt, B.; Taddeo, M.; Floridi, L. Artificial Intelligence and the 'Good Society': The US, EU, and UK approach. *Sci. Eng. Ethics* **2018**, *24*, 505–528. [CrossRef] [PubMed]
29. Computer laws revisited (Special issue). *Computer* **2013**, *46*, 12.
30. Gams, M. Intelligence in Information Society. In *Enabling Society with Information Technology*; Jin, Q., Li, J., Zhang, N., Cheng, J., Yu, C., Noguchi, S., Eds.; Springer: Tokyo, Japan, 2002. [CrossRef]
31. Gams, M.; Gu, I.Y.; Härmä, A.; Muñoz, A.; Tam, V. Artificial intelligence and ambient intelligence. *J. Ambient. Intell. Smart Environ.* **2019**, *11*, 71–86. [CrossRef]
32. Denning, P.J.; Lewis, T.G. Exponential Laws of Computing Growth. *Commun. ACM* **2017**, *60*, 54–65. [CrossRef]
33. Moore, G.E. Cramming more components onto integrated circuits. *Electronics* **1965**, *38*, 114–117. [CrossRef]
34. Markoff, J. The not-so-distant future of personal computing. *InfoWorld* **1993**, *15*, 48–50.
35. Borkar, S.; Chien, A.A. The Future of Microprocessors. *Commun. ACM* **2011**, *54*, 67. [CrossRef]
36. Bell, G. Bell's Law for the Birth and Death of Computer Classes. *Commun. ACM* **2008**, *51*, 86–94. [CrossRef]
37. Chip, W. Kryder's Law. *Sci. Am.* **2005**, *293*, 32–33. [CrossRef]
38. Antoniazzi, L. Digital preservation and the sustainability of film heritage. *Inf. Commun. Soc.* **2020**, 1–16. [CrossRef]
39. Salvadeo, P.A.; Veca, Á.C.; López, R.C. Historic behavior of the electronic technology: The Wave of Makimoto and Moore's Law in the Transistor's Age. In Proceedings of the 2012 VIII Southern Conference on Programmable Logic, Bento Goncalves, Brazil, 20–23 March 2012; pp. 1–5. [CrossRef]
40. Hruska, J. How Makimoto's Wave Explains the Tsunami of New AI Processors. Available online: https://www.extremetech.com/computing/287137-how-makimotos-wave-explains-the-tsunami-of-specialized-ai-processors-headed-for-market (accessed on 10 December 2020).
41. Hecht, J. Is Keck's Law Coming to an End? After Decades of Exponential Growth, Fiber-Optic Capacity May Be Facing a Plateau. Available online: https://spectrum.ieee.org/semiconductors/optoelectronics/is-kecks-law-coming-to-an-end (accessed on 10 December 2020).
42. Wilson, J.M. Computing, Communication, and Cognition. Three Laws That Define the Internet Society: Moore's, Gilder's, and Metcalfe's. Available online: http://www.jackmwilson.net/Entrepreneurship/Cases/Moores-Meltcalfes-Gilders-Law.pdf (accessed on 10 December 2020).
43. Koomey, J.; Berard, S.; Sanchez, M.; Wong, H. Implications of Historical Trends in the Electrical Efficiency of Computing. *IEEE Ann. Hist. Comput.* **2010**, *33*, 46–54. [CrossRef]
44. Dennard, R.H.; Gaensslen, F.H.; Yu, H.; Rideout, V.L.; Bassous, E.; LeBlanc, A.R. Design of ion-implanted MOSFET's with very small physical dimensions. *IEEE J. Solid-State Circuits* **1974**, *9*, 256–268. [CrossRef]
45. Schaller, B. The Origin, Nature, and Implications of 'Moore's Law. Available online: http://jimgray.azurewebsites.net/moore_law.html (accessed on 10 December 2020).
46. Hoshida, Y. Moore's Law Is Replaced by Neven's Law for Quantum Computing. Available online: https://community.hitachivantara.com/s/article/moores-law-is-replaced-by-nevens-law-for-quantum-computing (accessed on 10 December 2020).
47. Letzter, R. China Claims It's Achieved 'Quantum Supremacy' with the World's Fastest Quantum Computer. Available online: https://www.sciencealert.com/china-has-developed-the-fastest-and-most-powerful-quantum-computer-yet (accessed on 10 December 2020).
48. Rodgers, D.P. Improvements in multiprocessor system design. *ACM SIGARCH Comput. Archit. News* **1985**, *13*, 225–231. [CrossRef]
49. Gustafson, J.L. Reevaluating Amdahl's Law. *Commun. ACM* **1988**, *31*, 532–533. [CrossRef]
50. Grosch, H.R.J. High Speed Arithmetic: The Digital Computer as a Research Tool. *J. Opt. Soc. Am.* **1953**, *43*, 306–310. [CrossRef]
51. Raymond, E.S. *The Cathedral and the Bazaar*; O'Reilly Media: Sebastopol, CA, USA, 1999.
52. Wirth, N. A Plea for Lean Software. *Computer* **1995**, *28*, 64–68. [CrossRef]
53. Brooks, F.P., Jr. *The Mythical Man-Month*; Addison-Wesley: Boston, MA, USA, 1995.
54. Shapiro, C.; Varian, H.R. *Information Rules*; Harvard Business Press: Brighton, MA, USA, 1999.
55. Odlyzko, A.; Tilly, B. *A Refutation of Metcalfe's Law and a Better Estimate for the Value of Networks and Network Interconnections*; University of Minnesota: Minneapolis, MN, USA, 2005.
56. Tapscott, D. *The Digital Economy: Promise and Peril in the Age of Networked Intelligence*; McGraw-Hill: New York, NY, USA, 1997.
57. Bits and Atoms. Available online: https://www.wired.com/1995/01/negroponte-30/ (accessed on 10 December 2020).
58. Norberg, J. *Open: The Story of Human Progress*; Atlantic Books: London, UK, 2020.
59. Gross, B.M. *The Managing of Organizations: The Administrative Struggle*; Free Press of Glencoe: New York, NY, USA, 1964.
60. Challenge and Promise of E-Democracy. Available online: https://www.griffithreview.com/articles/challenge-and-promise-of-e-democracy/ (accessed on 10 December 2020).

61. Kidd, J. Are new media democratic? *Cult. Policy Crit. Manag. Res.* **2011**, *5*, 99–109.
62. Ownership Chart: The Big Six. Available online: http://files.meetup.com/17628282/Media-Big-Six.pdf (accessed on 10 December 2020).
63. American Views 2020: Trust, Media and Democracy. Available online: https://knightfoundation.org/reports/american-views-2020-trust-media-and-democracy/ (accessed on 10 December 2020).
64. Nagy, B.; Farmer, J.D.; Bui, Q.M.; Trancik, J.E. Statistical Basis for Predicting Tech-nological Progress. *PLoS ONE* **2013**, *8*, e52669. [CrossRef]
65. Wright, T.P. Factors affecting the costs of airplanes. *J. Aeronaut. Sci.* **1936**, *3*, 122–128. [CrossRef]
66. Ball, P. Moore's law is not just for computers. *Nature* **2013**. [CrossRef]
67. Bayus, B.L. An Analysis of Product Lifetimes in a Technologically Dynamic Industry. *Manag. Sci.* **1998**, *44*, 763–775. [CrossRef]
68. Unimaginable Output: Global Production of Transistors. Available online: https://www.darrinqualman.com/global-production-transistors/ (accessed on 10 December 2020).
69. Astronomers Have Found the Edge of the Milky Way at Last. Available online: https://www.sciencenews.org/article/astronomers-have-found-edge-milky-way-size (accessed on 10 December 2020).
70. How Many Solar Systems Are in Our Galaxy? Available online: https://spaceplace.nasa.gov/other-solar-systems/en (accessed on 10 December 2020).
71. Herculano-Houzel, S. The human brain in numbers: A linearly scaled-up primate brain. *Front. Hum. Neurosci.* **2009**, *3*, 31. [CrossRef] [PubMed]
72. 55th Anniversary of Moore's Law. Available online: https://www.infoq.com/news/2020/04/Moores-law-55/ (accessed on 10 December 2020).
73. Riding the S-Curve: The Global Uptake of Wind and Solar Power. Available online: https://www.uib.no/en/cet/127447/riding-s-curve-global-uptake-wind-and-solar-power (accessed on 10 December 2020).
74. Rotman, D. We're not prepared for the End of Moore's Law. Available online: https://www.technologyreview.com/2020/02/24/905789/were-not-prepared-for-the-end-of-moores-law (accessed on 10 December 2020).
75. Winston, P. *Artificial Intelligence*; Pearson: London, UK, 1992.
76. History of AI Winters. Available online: https://www.actuaries.digital/2018/09/05/history-of-ai-winters/ (accessed on 10 December 2020).
77. Moloi, T.; Marwala, T. *Artificial Intelligence in Economics and Finance Theories*; Springer: Berlin/Heidelberg, Germany, 2020.
78. IJCAI Conference. 2017. Available online: https://ijcai-17.org (accessed on 4 February 2021).
79. Jug, J.; Kolenik, T.; Ofner, A.; Farkas, I. Computational model of enactive visuospa-tial mental imagery using saccadic perceptual actions. *Cogn. Syst. Res.* **2018**, *49*, 157–177. [CrossRef]
80. IEEE Computational Intelligence Society. Available online: https://cis.ieee.org/ (accessed on 10 December 2020).
81. Gjoreski, M.; Janko, V.; Slapničar, G.; Mlakar, M.; Reščič, N.; Bizjak, J.; Drobnič, V.; Marinko, M.; Mlakar, N.; Luštrek, M.; et al. Classical and deep learning methods for recognizing human activities and modes of transportation with smartphone sensors. *Inf. Fusion* **2020**, *62*, 47–62. [CrossRef]
82. Yun, Y.; Gu, I.Y.H. Riemannian Manifold-Valued Part-Based Features and Geodesic-Induced Kernel Machine for Human Activity Classification Dedicated to Assisted Living. *Comput. Vis. Image Underst.* **2017**, *161*, 65–76. [CrossRef]
83. The 4 Deep Learning Breakthroughs You Should Know about. Available online: https://towardsdatascience.com/the-5-deep-learning-breakthroughs-you-should-know-about-df27674ccdf2 (accessed on 10 December 2020).
84. Zhang, X.; Tian, Q.; Wang, L.; Liu, Y.; Li, B.; Liang, Z.; Gao, P.; Zheng, K.; Zhao, B.; Lu, H. Radiomics Strategy for Molecular Subtype Stratification of Lower-Grade Glioma: Detecting IDH and TP53 Mutations Based on Multimodal MRI. *J. Magn. Reason. Imaging* **2018**, *48*, 916–926. [CrossRef]
85. Ge, C.; Gu, I.Y.H.; Jakola, A.S.; Yang, J. Deep Learning and Multi-Sensor Fusion for Glioma Classification using Multistream 2D Convolutional Networks. In Proceedings of the 40th Annual International Conference of the IEEE Engineering in Medicine and Bi-ology Society (EMBC'18), Honolulu, HI, USA, 18–21 July 2018; pp. 5894–5897.
86. Chang, K.; Bai, H.X.; Zhou, H.; Su, C.; Bi, W.L.; Agbodza, E.; Kavouridis, V.K.; Senders, J.T.; Boaro, A.; Beers, A.; et al. Residual Convolutional Neural Network for the Determination ofIDHStatus in Low- and High-Grade Gliomas from MR Imaging. *Clin. Cancer Res.* **2018**, *24*, 1073–1081. [CrossRef]
87. Eye Scans to Detect Cancer and Alzheimer's Disease. Available online: https://spectrum.ieee.org/the-human-os/biomedical/diagnostics/eye-scans-to-detect-cancer-and-alzheimers-disease (accessed on 10 December 2020).
88. Sarraf, S.; Tofighi, G. DeepAD: Alzheimer's disease classification via deep convolutional neural networks using MRI and fMRI. *bioRxiv* **2016**. [CrossRef]
89. Bäckström, K.; Nazari, M.; Gu, I.Y.H.; Jakola, A.S. An efficient 3D deep convolutional network for Alzheimer's disease diagnosis using MR images. In Proceedings of the 2018 IEEE 15th International Symposium on Biomedical Imaging, Washington, DC, USA, 4–7 April 2018.
90. Pejović, V.; Gjoreski, M.; Anderson, C.; David, K.; Luštrek, M. Toward Cognitive Load Inference for Attention Management in Ubiquitous Systems. *IEEE Pervasive Comput.* **2020**, *19*, 35–45. [CrossRef]
91. Gjoreski, M.; Kolenik, T.; Knez, T.; Luštrek, M.; Gams, M.; Gjoreski, H.; Pejović, V. Datasets for Cognitive Load Inference Using Wearable Sensors and Psychological Traits. *Appl. Sci.* **2020**, *10*, 3843. [CrossRef]

92. Kolenik, T.; Gams, M. Persuasive Technology for Mental Health: One Step Closer to (Mental Health Care) Equality? *IEEE Technol. Soc. Mag.* **2020**. [CrossRef]
93. Kurzweil, R. *The Singularity Is Near: When Humans Transcend Biology*; Penguin Books: London, UK, 2006.
94. Bostrom, N. *Superintelligence—Paths, Dangers, Strategies*; Oxford University Press: Oxford, UK, 2014.
95. Kolenik, T. Seeking after the Glitter of Intelligence in the Base Metal of Computing: The Scope and Limits of Computational Models in Researching Cognitive Phenomena. *Interdiscip. Descr. Complex Syst.* **2018**, *16*, 545–557. [CrossRef]
96. Beard, J.M. Autonomous Weapons and Human Responsibilities. *Georget. J. Int. Law* **2014**, *45*, 617.
97. Will the Latest AI Kill Coding? Available online: https://towardsdatascience.com/will-gpt-3-kill-coding-630e4518c04d (accessed on 10 December 2020).

Article

Ultra-Short Window Length and Feature Importance Analysis for Cognitive Load Detection from Wearable Sensors

Jaakko Tervonen [1,*], Kati Pettersson [1] and Jani Mäntyjärvi [2]

1. VTT Technical Research Centre of Finland, 02044 Espoo, Finland; kati.pettersson@vtt.fi
2. VTT Technical Research Centre of Finland, 90571 Oulu, Finland; jani.mantyjarvi@vtt.fi
* Correspondence: jaakko.tervonen@vtt.fi

Abstract: Human cognitive capabilities are under constant pressure in the modern information society. Cognitive load detection would be beneficial in several applications of human–computer interaction, including attention management and user interface adaptation. However, current research into accurate and real-time biosignal-based cognitive load detection lacks understanding of the optimal and minimal window length in data segmentation which would allow for more timely, continuous state detection. This study presents a comparative analysis of ultra-short (30 s or less) window lengths in cognitive load detection with a wearable device. Heart rate, heart rate variability, galvanic skin response, and skin temperature features are extracted at six different window lengths and used to train an Extreme Gradient Boosting classifier to detect between cognitive load and rest. A 25 s window showed the highest accury (67.6%), which is similar to earlier studies using the same dataset. Overall, model accuracy tended to decrease as the window length decreased, and lowest performance (60.0%) was observed with a 5 s window. The contribution of different physiological features to the classification performance and the most useful features that react in short windows are also discussed. The analysis provides a promising basis for future real-time applications with wearable sensors.

Citation: Tervonen, J.; Pettersson, K.; Mäntyjärvi, J. Ultra-Short Window Length and Feature Importance Analysis for Cognitive Load Detection from Wearable Sensors. *Electronics* **2021**, *10*, 613. https://doi.org/10.3390/electronics10050613

Academic Editor: Maysam Abbod

Received: 17 December 2020
Accepted: 3 March 2021
Published: 6 March 2021

Publisher's Note: MDPI stays neutral with regard to jurisdictional claims in published maps and institutional affiliations.

Copyright: © 2021 by the authors. Licensee MDPI, Basel, Switzerland. This article is an open access article distributed under the terms and conditions of the Creative Commons Attribution (CC BY) license (https://creativecommons.org/licenses/by/4.0/).

Keywords: machine learning; affective computing; cognitive load; psychophysiology; supervised learning

1. Introduction

In the near future, unobtrusive, reliable, and affordable wearable sensors will enable cognitive state estimation of a person in real-time. The cognitive state, i.e., a person's overall capacity and readiness to meet everyday situations, is affected by various conditions such as sleep deprivation [1,2], acute stress [3,4], and cognitive load [5] and thus cognitive state estimation would be beneficial in many application areas, e.g., transportation, industry, rehabilitation, and education.

In many working environments, the modern technology such as human computer interaction (HCI) systems impose high cognitive demands for humans, thus increasing the cognitive load of a person [6]. Real-time assessment of a person's cognitive load could be used to identify overload situations where the probability of error is increased. Further, in the near future, HCI and cyber-physical systems could use the information to optimize user interface content and interactions to match the imposing workload with the prevailing cognitive capacity of the user. However, this would require seamless operation between the HCI system and the users, meaning accurate and real-time (with minimal delay) assessment of the cognitive load.

Humans respond to external stimuli by adjusting nervous system functions, which causes physiological reactions that can be detected from different type of biosignals. The autonomic nervous system (ANS) is one of the major neural pathways activated by stress [7]: the sympathetic branch of the ANS prepares body for an emergency while the parasympathetic branch facilitates recovery [8]. An increase in the heart rate (HR) reflects the sympathetic nervous system (SNS) activation while parameters derived from the heart rate variability (HRV) parameters can capture variations both in the SNS and parasympathetic nervous system (PNS) activations [9]. Galvanic skin response (GSR) reflects the activity in the sweat glands, which are solely connected to the SNS. Therefore, the GSR is considered to be an undisturbed measure of SNS activation [8]. In addition, in acute stress the SNS triggers peripheral vasoconstriction which reduces the flow in the blood vessels and reduces the skin temperature (ST) [10]. However, the after a short delay the blood flow recovers resulting in delayed skin warming [9,11,12].

The changes in the cognitive load are also reflected in various biosignals that can be measured by using biosensors, e.g., wearable devices [13,14]. For instance, increasing task difficulty (or cognitive load) and acute stress increases the HR and breathing rate [15], ST [11] as well as GSR [16] and decreases the HRV [17], number of eye movements [18], and increases the blink rate [19,20].

Real-time cognitive load estimation means processing a stream of biosignals with minimal latency. Research on affective, or cognitive state/load, detection systems has focused mainly on state recognition methodology and optimizing the used sensor set (see, e.g., [21]). To achieve real-time or continuous monitoring of the cognitive state/load, the segmentation part (i.e., selection of used window length) of the state detection pipeline has received little attention and it requires further research.

The cognitive load is estimated from various biosignals and each of these signals has its own characteristics. For instance, HR could be considered as a periodic signal, whereas some other biosignals, such as eye movements and GSR reactivity, have a bursty nature and are more linked to the stimulus or task at hand. Further, the level of some slow-acting signals, skin temperature and the tonic component of the GSR signal, may increase or decrease during a cognitive load (e.g., due to changes in alertness). Thus, the varying nature of the biosignals sets limits to the window lengths: the length must be long enough to include sufficient variation and periods for the periodic signals but short enough that bursty events do not average out.

In recent studies the window lengths have varied (see Table 1) from 1 s to 360 s. In most studies, the window lengths have been selected based on the physiology, task duration, or previous studies. However, the literature on ultra-short windows (<60 s, especially <30 s), e.g., in HR and HRV analyses is rather limited (see the review by Shaffer and Ginsberg [22]) and therefore, there may not be theoretical limits for the physiological features used in real-time/continuous cognitive state estimation. In addition, there are few studies where the effect of window length to classification accuracy has been studied (see Table 1) and even those have mainly used windows with length of 30 s or more.

Table 1. Previous studies in the affective computing domain with an emphasis on state detection based on physiological variables conducted mostly in constrained or laboratory environments.

Study	Signals (Sampling Rate in Hz)	Window Lengths Used	Overlap	Optimal Window Length	Model	Classification Performance
Cognitive state detection						
[9]	GSR (1000), ST (1000), HRV **	30–300 s	0–90%	30 s/60 s	LDA, kNN, QDA, SVM	97% accuracy (binary)
[14]	ST (1), GSR (1), HR (1), HRV (1)	30 s	-	-	Bagging, XGB	68% and 82% accuracy (binary, two datasets)
[23]	EOG (250)	1–10 s	-	10 s	LDA	87% accuracy (binary)
[20]	HR **, HRV **, EOG (1000)	45 s	15 s	-	XGB, SVM	86% accuracy (three classes) 97% accuracy (binary)
Stress detection						
[24]	ST (4), GSR (4), HR (1), HRV *	30–360 s	5–275 s	300 s	SVM	73% accuracy (three classes)
[25]	ST (32), GSR (32), HR **, HRV **, RESP (32)	30 s	29 s	-	dBN	85% mean of sensitivity and specificity (binary)
[26]	HRV **	30–300 s	-	300 s	kNN	94% accuracy
[27]	GSR (32), HRV **	50 s	30 s	-	SOM	79% mean of sensitivity and specificity (binary)
Emotion detection						
[28]^F	ACC (NA), GSR (4), HR (NA)	60–300 s	-	-	DT, BN	51% error rate on detecting arousal
[29]	EEG (256), HRV **	90 s	-	-	SVM	75% and 82% accuracy (valence and arousal, binary)
[30]	HRV **	90 s	-	-	SVM	71% accuracy (positive or negative emotion)
[31]	EEG (256)	1 s	-	-	LDA	73% accuracy (positive or negative emotion)
Stress/emotion detection						
[32]	ACC (32), ST (4), GSR (4), HR (1), HRV *	15–120 s	0.25 s window slide	120 s	LDA	84% accuracy (binary)
[33]	ACC (32), ST (4), GSR (4), HR (1), HRV *	60 s	0.25 s window slide	-	RF	76% accuracy (three classes) 88% accuracy (binary)
Odor pleasantness classification						
[34]	EEG (250), HRV **	6 s	-	-	LDA	0.46 Cohen's kappa

^{F} Field study. * Interbeat intervals used to derive HRV were obtained on-device from blood volume pulse signal sampled at 64 Hz; ** Interbeat intervals used to derive HR/HRV were obtained from an electrocardiogram sampled at 200 Hz [30], 250 Hz [25,27,29], 256 Hz [34], and 1000 Hz [9,20]. Abbreviations: ACC: acceleration, (d)BN: (dynamic) Bayesian network, DT: decision tree, EEG: electroencephalogram, EOG: electro-oculogram, GSR: galvanic skin response, HR: heart rate, HRV: heart rate variability, kNN: k-nearest neighbors, LDA: linear discriminant analysis, NA: not available, QDA: quadratic discriminant analysis, RESP: respiration, RF: random forest, SOM: self-organizing map, ST: skin temperature, SVM: support vector machine, XGB: extreme gradient boosting.

Healey et al. [28] attempted emotion detection in a field study in windows of 60 s, 180 s, and 300 s, but the best window length was not reported since each one showed poor performance. Gjoreski et al. [24] experimented with window lengths between 30 s and 360 s in a laboratory study of stress detection, and selected the 300 s window for a continuation study with field data. Anusha et al. [9] found that a 30 s window performed the best for ST, and a 60 s window for GSR in cognitive state detection; however, window length experiments were not conducted for HRV. Marshall [23] detected the cognitive state based on eye movements and found that a 10 s window provided highest detection accuracy. Siirtola [32] studied stress detection in a laboratory with window lengths between 15 s and 120 s. It was found that whereas the 120 s window performed the best, a 15 s window performed better than a 30 s window and almost the same as a 60 s window, which shows that window lengths shorter than 30 s have the potential to perform well despite containing less data than longer windows.

In a related context, Kroupi et al. [34] detected odor pleasantness in 6 s windows based on electroencephalogram and HRV measurements. Moreover, Kreibig [8] reports on multiple studies using shorter than 30 s averaging periods for physiological responses. However, the goal in those studies was to observe the effects emotions have on functions of the autonomic nervous system, rather than classifying between emotional/cognitive states based on those effects.

Thus, the existing research on cognitive state recognition has not focused on the segmentation part of the state detection pipeline. Even when experiments with different window lengths have been conducted, they have focused on rather long window lengths, despite the fact that shorter window lengths have been considered in related contexts. The novelty in this study is on performing a systematic comparison of ultra-short windows (30 s or less) in terms of the classification performance for cognitive load detection. An analysis of the contribution of different features is also presented, and the variation of the most useful features between tasks is discussed. Further, individual differences related to the optimal window length and feature variation between the study subjects as well as the effect of optimizing classifier hyperparameters are studied.

2. Materials & Methods

2.1. Dataset

The CogLoad dataset from [14] was used in this study. The dataset includes 23 participants (7 females, mean age 29.5 years with a standard deviation of 10.1 years) who solved cognitive tasks of varying difficulty. In the first part, the participants solved N-back tasks, i.e., 2-back and 3-back tasks, with a three-minute rest after each of them, and answered questions to determine their personality. In the second part, six elementary cognitive tasks (ECT) each with three difficulty levels were presented: the Gestalt Completion test (GC), the Hidden Pattern test (HP), Finding A's test (FA), Number Comparison test (NC), Pursuit test (PT), and Scattered X's test (SX), with a rest period between them. After each task, the participants were asked to fill in the NASA-TLX questionnaire to determine their subjective cognitive load, however, those questionnaires were not utilized here. Further details on the study protocol and tasks can be found in [14].

While doing the tasks, the participants' physiological response was measured with a wrist device (Microsoft Band). The measurements included the HR, R-to-R intervals (RR), GSR, ST and 3-axis acceleration, which was not used in this study. The open-sourced dataset contains the data re-sampled to a frequency of 1 Hz. However, the HR and RR were derived on-device from an optical sensor and the raw measurements used to obtain those two signals were not available. Thus, the rate at which the HR and RR were measured was truly not constant but dynamic, and depended on when the heartbeats occurred.

Figure 1 depicts the steps taken in analyzing the dataset and evaluating the results.

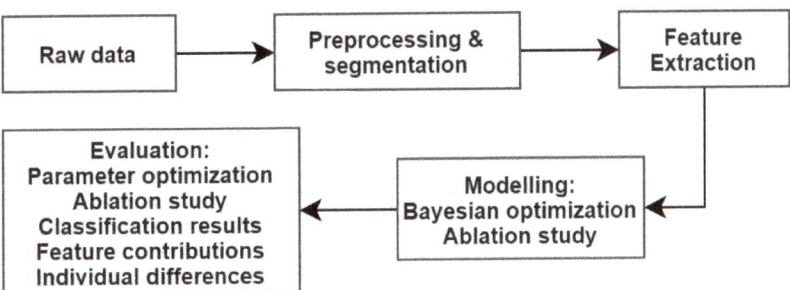

Figure 1. Pipeline followed in data processing and evaluating the results.

2.2. Data Preprocessing, Segmentation and Feature Extraction

The main focus in this study was on evaluating the classification performance of cognitive load at different window lengths of less than 30 s in duration. Window lengths selected were 5 s, 10 s, 15 s, 20 s, 25 s, and 30 s in duration, and a 50% window overlap was employed to increase the amount of data.

The time taken by the participants to complete the tasks varied between 18 s and 190 s. If a task lasted for a shorter time than window length, that single task was removed from the experiment with the specific window length to make sure that a shorter actual task length would not skew the results; approximately 4% of all tasks were completed in less than 30 s. In addition, it was noted that sometimes the data had been filled by carrying the last observation forward, i.e., a signal was constant for a period of time. As many features could not be calculated if there was not enough variation, segments with less than 25% unique values in the RR-, HR-, or GSR-signal were removed.

Next, features were extracted at each window length. According to [21,35], features that are usually extracted from the signals used here contain the statistics of each signal, heart rate variability from the RR-signal, and skin conductance response analysis for the GSR signal.

In this study, the statistical features of the RR, HR, GSR, and ST and their first and second derivatives were computed. The statistical features included the mean, standard deviation, minimum, maximum, difference between minimum and maximum, lower and upper quartile, interquartile range, and coefficient of variation.

A skin conductance response (SCR) analysis was conducted for the GSR signal to extract additional features. Like in the original paper using the same dataset [14], the signal was first preprocessed with a sliding mean filter, and then fast-acting (phasic) and slow-acting (tonic) components were extracted. Normally, the SCR analysis is used especially to extract features from the phasic component and SCR peaks [21,35]. In this analysis, however, it often happened that a segment did not contain any SCR peaks, especially with shorter window lengths. Therefore, the features extracted from the phasic component included the number of SCR peaks and the statistics (mean, standard deviation, median, lower and upper quartile, minimum and maximum) of its first and second derivative, and the total time the first derivative of the phasic component was positive (rise-time) and negative (descend-time). The features extracted from the tonic component included its mean, standard deviation, minimum, maximum, ratio of maximum and minimum, and its correlation with time.

Additionally, heart rate variability (HRV) features were extracted from the R-to-R intervals. Following [22,36], the HRV features extracted included the mean, median, and range of normal-to-normal intervals, standard deviation of normal-to-normal intervals (SDNN) and successive differences, percentage and number of normal-to-normal intervals differing by more than 20 ms and 50 ms, root mean square of successive differences (RMSSD), ratio of SDNN and mean normal-to-normal intervals (CVNNI), ratio of RMSSD and mean normal-to-normal intervals (CVSD), power in very low, low and high frequency bands, total power, ratio of low and high frequency power, normalised low and high frequency power, triangular index, (modified) cardiac sympathetic index, cardiac vagal index, and Poincaré plot indices SD1, SD2, and SD1/SD2.

A total of 157 features were extracted and they are listed in Table 2. Afterwards, a sanity check was conducted for the features computed. Some features had a significant amount of missing or infinite values, or showed little variation. Thus, features with missing values, infinite values, or variance below 0.01 were removed for each window length. The number of remaining features was 93 at window lengths from 20 s to 30 s, 91 at window lengths of 10 s and 15 s, and 82 at a window length of 5 s.

Table 2. List of computed features. Abbreviations used later in text and in figures are in parenthesis.

Category	Features
Statistical features for each signal (d0) and their first (d1) and second (d2) derivatives	mean, standard deviation (std), minimum (min), maximum (max), difference between minimum and maximum (range), lower (lq) and upper quartile (uq), interquartile range (iqr), and coefficient of variation (cv)
Heart rate variability (HRV)	mean, median, and range of normal-to-normal intervals (mean, median, range nni), standard deviation of normal-to-normal intervals (sdnn) and successive difference (sdsd), root mean square of successive differences (rmssd), ratio of sdnn and mean nni (cvnni), ratio of rmssd and mean nni (cvsd), signal power in very low (vlf), low (lf), and high (hf) frequency bands, total power, ratio of lf and hf, normalised lf and hf (lfnu, hfnu), triangular index, cardiac sympathetic index (csi), modified csi, cardiac vagal index (cvi), and Poincaré plot features (SD1, SD2, ratio of SD1 and SD2)
Skin conductance response (SCR)	phasic component: number of peaks (npeaks), time first derivative was positive (risetime) and negative (dectime), and mean, std, median, lq, and uq of first (diff1) and second (diff2) derivatives tonic component (scl): mean, std, min, max, ratio of max and min (slope), and its correlation with time (corrwithtime)

According to the criteria stated above, features that were removed most often were the HRV parameters CVSD and CVNNI, coefficient of variation of the HR and its derivatives, statistical features of the derivatives of the phasic component of the GSR signal, standard deviation of the tonic component of the GSR signal, and statistical features of the derivatives of the ST signal across the different window lengths. Additionally, several HRV parameters were removed from the shortest window length.

The features were normalized using within-subject standardization, meaning that each feature was transformed by subtracting its mean and dividing by its standard deviation separately for each participant. Person-specific standardization was conducted instead of person-independent standardization, since it has shown improved performance in earlier work in similar contexts [14,20,37].

2.3. Model and Experimental Protocol

The classification task was formulated as binary classification between a cognitive load and a rest class. All data segments during a cognitive task (N-back task or one of the six ECTs) were annotated as a cognitive load, and all the rest periods were annotated as rests. The segments during which the participant answered the questionnaires were removed from the data. The number of instances in both classes at each window length are reported in Table 3. Because the number of samples in each class is reasonably balanced (approximately 46% rests and 54% cognitive loads for each window length), the classification performance was assessed in terms of accuracy, the percentage of correctly classified samples.

Table 3. Number of instances in cognitive load and rest classes at each window length.

	30 s	25 s	20 s	15 s	10 s	5 s
Cognitive load	1654	2110	2840	4055	6336	12,763
Rest	1406	1791	2407	3437	5393	10,937

Extreme Gradient Boosting [38] (XGB) was selected as the classification model for the following reasons: the classifiers from the random forest family and the boosting method have been shown to be strong classifiers [39], XGB has shown good performance earlier in similar contexts [14,20] and because XGB is computationally efficient and scales to very large datasets [38]. XGB is an ensemble of decision trees, each of which splits the data hierarchically, aiming to contain data originating from a single class in each leaf node. XGB uses the gradient descent algorithm to construct the trees sequentially so that each subsequent tree attempts to fix the errors made by preceding trees.

Specifically, XGB aims to minimize the regularized objective function

$$\mathcal{L}(\hat{y}) = \sum_{i=1}^{n} l(\hat{y}_i, y_i) + \sum_{k=1}^{K} \Omega(f_k), \quad (1)$$

where $\Omega(f) = \gamma T + \frac{1}{2}\lambda||w||^2$. Here, n is the number of observations, l is a differentiable convex loss function measuring the difference between the prediction $\hat{y}_i$ and the target y_i, K is the number of classification trees, f_k are classification tree functions, Ω is a regularization function penalizing the complexity of the model, T is the number of leaves in a tree, γ and λ are regularization parameters and w are leaf weights. Assume that I_L and I_R are the instance sets of left and right nodes after a split. Then, letting $I = I_L \cup I_R$, the loss reduction after the split is given by

$$\mathcal{L}_{split} = \frac{1}{2}\left[\frac{(\sum_{i \in I_L} g_i)^2}{\sum_{i \in I_L} h_i + \lambda} + \frac{(\sum_{i \in I_R} g_i)^2}{\sum_{i \in I_R} h_i + \lambda} - \frac{(\sum_{i \in I} g_i)^2}{\sum_{i \in I} h_i + \lambda}\right] - \gamma, \quad (2)$$

where $g_i = \partial_{\hat{y}^{(t-1)}} l(y_i, \hat{y}^{(t-1)})$ and $h_i = \partial^2_{\hat{y}^{(t-1)}} l(y_i, \hat{y}^{(t-1)})$ are the first and the second order gradient statistics of the loss function at the t-th iteration. In practice, the formula above is used for evaluating the split candidates, and the splits are found either with an exact or approximate greedy algorithm that are implemented in [38]. So, at each boosting iteration t, a classification tree f_t whose splits are found using a greedy algorithm with Equation (2) and that most improves the model according to Equation (1) is added to the model.

The performance of the XGB model depends on its hyperparameters that describe the structure of each tree and that affect the convergence of the loss function. The hyperparameters were optimized using Bayesian optimization (see Section 2.4) but an ablation study using default hyperparameters was also conducted. The following hyperparameters were optimized:

- *max_depth*: the maximum depth of each tree
- *n_estimators*: the number of estimators in the model
- *reg_alpha*: L1 regularization term
- *reg_lambda*: L2 regularization term
- *subsample*: the ratio training instances used for each boosting iteration
- *learning_rate*: the step size shrinkage used in each update to prevent overfitting
- *gamma*: the minimum loss reduction required to make a further split on a leaf node of the tree
- *colsample_bytree*: the ratio of the number of features used to create each tree
- *colsample_bynode*: the ratio of the number of features used at each node (split)
- *colsample_bylevel*: the ratio of the number of features used at each tree level

The dataset was published with participants divided into training and testing sets, with 18 subjects for training and 5 for testing. However, the ablation study without Bayesian optimization showed significantly higher performance for the test subjects than the training subjects even though the model had not seen the data of the test subjects. Therefore, test subjects' data appeared to be different from the data of the training subjects. So, instead of using the fixed train-test-split the dataset came with, it was decided to use cross-validation with both the training and testing subjects in a single pool to validate the modelling results with Bayesian optimization. Individual differences are further elaborated in Section 3.4.

In general, leave-one-subject-out (LOSO) validation is recommended in the affective computing domain [21], meaning that each subject is left out in turn for testing and the rest of the data is used for training the model. Moreover, when tuning hyperparameters, an internal validation with training data is required to make sure that the best hyperparameters are selected according to the validation performance, and not the testing performance.

Instead of LOSO validation, it was decided to use the leave-two-subjects-out (LTSO) validation method when tuning hyperparameters because the process is computationally intensive, especially since it had to be completed for each window size separately, and because the number of participants in the dataset was relatively large (LOSO validation would correspond to 23-fold cross-validation). So, each hyperparameter configuration was evaluated with data of two randomly selected subjects left out for testing, with internal leave-two-subjects-out validation to select the hyperparameters. This had the effect of approximately halving the computation time during the Bayesian optimization compared to using LOSO validation. However, to comply with earlier research the final results are also reported with LOSO validation for the best hyperparameter configuration.

2.4. Hyperparameter Optimization

Bayesian optimization is a derivative-free search strategy for the global optimization of functions that are expensive to evaluate. The algorithm starts by setting a prior distribution over the parameters to optimize and evaluating the function (here, a function value refers to LTSO validation accuracy with the XGB model) a certain number of times on parameter values sampled from the prior distribution. Then, for a set number of iterations, posterior distributions of each parameter over the function are updated using all the available data, the values maximizing an acquisition function over the current posteriors are sampled, and the function is evaluated using those values. For more details on the algorithm we refer to [40].

Table 4 lists the hyperparameters that were optimized and the priors used for each parameter. Overall, non-informative priors were employed, and the prior distributions were either discrete uniform distributions on a given interval and step size (parameters *max_depth* and *n_estimators*), continuous uniform distributions on a given interval (parameters *reg_alpha*, *reg_lambda*, and *learning_rate*), or a random choice between a constant or a number sampled from a continuous uniform distribution on a given interval (parameters *subsample*, *gamma*, *colsample_bytree*, *colsample_bynode*, and *colsample_bylevel*).

Optimization was continued for a total of 300 iterations at each window length. The number of iterations was selected experimentally. As seen in Section 3.1, the performance improved little after 100–150 iterations. Thus, the procedure was continued for twice that long since it would have been unlikely that scores would improve much after that.

Table 4. Hyperparameters optimized for the XGB model, and their prior distributions using the *hyperopt* syntax.

Hyperparameter	Prior
max_depth	hp.quniform('max_depth-xg', 2, 12, 1)
n_estimators	hp.quniform('n_estimators-xg', 20, 250, 10)
reg_alpha	hp.uniform('reg_alpha-xg', 0, 1)
reg_lambda	hp.uniform('reg_lambda-xg', 0, 1)
subsample	hp.choice('subsample', [1, hp.uniform('subsample-xg', 0.7, 1)])
learning_rate	hp.uniform('learning_rate-xg', 0.01, 0.5)
gamma	hp.choice('gamma', [0, hp.uniform('gamma-xg', 0, 0.05)])
colsample_bytree	hp.choice('colsample_bytree', [1, hp.uniform('colsample_bytree-xg', 0.7, 1)])
colsample_bynode	hp.choice('colsample_bynode', [1, hp.uniform('colsample_bynode-xg', 0.7, 1)])
colsample_bylevel	hp.choice('colsample_bylevel', [1, hp.uniform('colsample_bylevel-xg', 0.7, 1)])

2.5. Statistical Tests

Statistical tests were conducted to determine (1) whether there was a statistically significant difference in the classification performance between the different window lengths, and (2) whether hyperparameter optimization provided statistically significantly more accurate classification (effect of ablating Bayesian optimization). First, subject-by-subject accuracy was obtained for each subject by training the model with all the other subjects in training data. Then, paired *t*-tests were employed for both scenarios. Pairs were formed from the accuracies observed for each subject at (1) two different window lengths and (2) for optimized and default hyperparameters. A *t*-test was selected since the subject-by-subject accuracies were normally distributed. In all tests, the Benjamini-Hochberg correction was used to control the false discovery rate, with probability of type I error set to 0.05.

2.6. Computational Tools

The analysis was completed using the Python programming language. The libraries employed were *scikit-learn* (preprocessing and cross-validation implementation) [41], *xgboost* (implementation of XGB model) [38], *hrv-analysis* (calculating HRV features) [42], *neurokit2* (signal processing and SCR analysis) [43], and *hyperopt* (Bayesian optimization) [44].

3. Results

3.1. Parameter Optimization

Figure 2 shows the evolution of the hyperparameter optimization. It is evident from the figure that most improvement took place within the first one hundred iterations for each window length, with only minor improvements afterwards. Overall, 30 s and 25 s window lengths performed similarly, and the accuracy decreased as the window length decreased further.

Figure 3 displays the posterior distributions of the *max_depth* parameter which describes the depth of each tree in the XGB model. The distribution of each window length is similar to the Gamma distribution truncated between 2 and 12 (since the prior was truncated between 2 and 12). Most of the probability mass was located between the depths from 2 to 6, and the best value found (black vertical lines) are located at depths 2 and 3 for all window lengths except for 25 s, which achieved its best performance at a maximum depth of 6.

Similar figures of posterior distributions for the rest of the hyperparameters are available as supplementary material, together with a table of all the tested hyperparameter configurations and their performance for each window length.

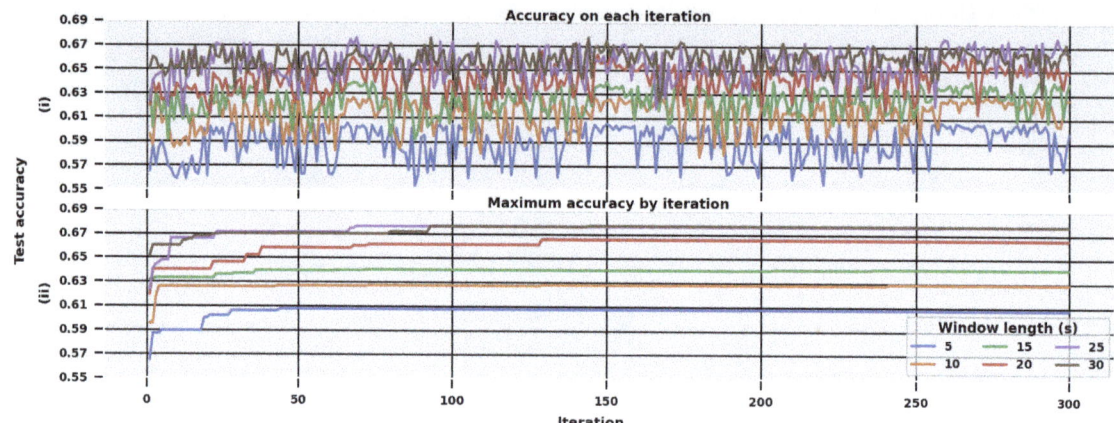

Figure 2. Progress of the Bayesian parameter optimization, (**i**) test accuracy obtained at each iteration with leave-two-out validation (**top** panel), and (**ii**) the cumulative maximum accuracy obtained by each iteration (**bottom** panel).

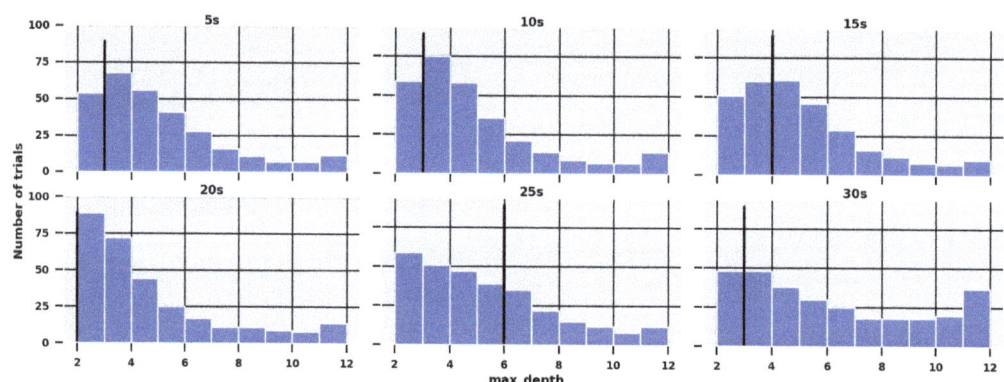

Figure 3. Posteriors of the *max_depth* parameter, describing the depth of each tree in the ensemble, after 300 Bayesian optimization iterations at each window length. Vertical lines denote the best value found.

3.2. Classification Results

Table 5 presents the classification accuracies obtained for each window length with the LTSO validation used during the Bayesian optimization, and with the LOSO validation with optimized and default model parameters. The results for statistical tests for the performance with default and optimized hyperparameters, and the different window lengths, are shown in Tables 6 and 7, respectively.

There were no differences in the mean classification performance between the LTSO and LOSO validation strategies (less than 1% difference at each window length), but LTSO validation seemed to underestimate standard deviations compared to LOSO validation. Since standard deviations were calculated from the overall accuracy and not the subject-specific accuracy across the folds, the distribution was drawn towards the mean when there were two subjects in a fold, which resulted in a lower standard deviation. LOSO validated measures using the optimized hyperparameters yielded a statistically significantly higher classification accuracy than with the default parameters. The difference between the default and optimized parameters was 2–3% at higher window lengths (20–30 s) and 3–4% at lower window lengths (5–15 s).

Table 5. The classification accuracy (%) for each window length, with leave-two-out validation with optimized parameters (LTSOopt), and leave-one-out validation with optimized parameters (LOSOopt) and default model parameters (LOSOdef). Standard deviations in parenthesis.

	30 s	25 s	20 s	15 s	10 s	5 s
LTSOopt	66.9 (4.9)	67.4 (4.5)	66.1 (5.2)	64.1 (4.1)	62.6 (4.3)	60.8 (3.5)
LOSOopt	67.2 (9.0)	67.6 (8.6)	65.4 (8.0)	63.6 (7.7)	62.2 (7.6)	60.0 (6.4)
LOSOdef	65.0 (7.5)	64.5 (8.3)	63.5 (6.6)	60.1 (6.6)	58.5 (6.5)	56.4 (5.7)

Table 6. Results for the Benjamini-Hochberg corrected paired t-tests between the default and optimized hyperparameters with LOSO validation.

30 s		25 s		20 s		15 s		10 s		5 s	
t	p	t	p	t	p	t	p	t	p	t	p
−2.40	0.03	−4.13	<0.001	−2.62	0.02	−3.63	<0.001	−5.09	<0.001	−6.12	<0.001

Table 7. Results for the Benjamini-Hochberg corrected paired t-tests between accuracies for different window lengths.

	30 s		25 s		20 s		15 s		10 s	
	t	p	t	p	t	p	t	p	t	p
25 s	−0.64	0.53								
20 s	1.95	0.07	2.38	0.03						
15 s	3.48	0.003	3.82	0.002	3.25	0.005				
10 s	4.2	0.001	4.26	0.001	4.78	<0.001	3.3	0.004		
5 s	5.58	<0.001	5.69	<0.001	7.19	<0.001	6.92	<0.001	5.84	<0.001

According to Table 7, there was no statistically significant difference when a window length of 30 s was compared to window lengths of 25 s and 20 s, but all other tests showed significant differences (at significance level $\alpha = 0.05$). Thus, apart from the longest window length, a shorter window always resulted in weaker classification performance.

3.3. Feature Contribution Results

The feature importance for the different window lengths was assessed as the normalized total reduction of the Gini impurity brought by that feature in the optimized XGB model. This is visualized in Figure 4. Regardless of the window length, the most important features seem to be related to the heart rate variability and R-to-R intervals statistics, and to the statistics of the derivatives of the GSR signal.

Partial dependence plots of a few features selected from the top-20 with a window length of 25 s are shown in Figure 5. The figure for the 25 s window features was selected since that option provided the highest classification performance, and the six features included were selected from the top-20 features so that each feature category or signal was included. Partial dependence plots display the effect that each feature has on the classification outcome when all other variables are kept constant. The scale of each feature is relative to each subject's feature value, since the features were normalized person-specifically.

For both HRV variables included, it seems that there is a sudden drop in partial dependence when the feature value approaches zero, i.e., the subject-specific mean, which suggests that both features were used near or at the root of the tree to split the whole of the data in half. The higher RR (rr_d0_lq), higher ST (st_d0_uq), and lower range of the second derivative of the GSR (gsr_d2_range) seem to be related to a higher chance of being classified in the cognitive load class. The mean HR shows a small effect on the outcome overall, but it seems that a higher HR is related to a lower chance of being classified in a cognitive load class.

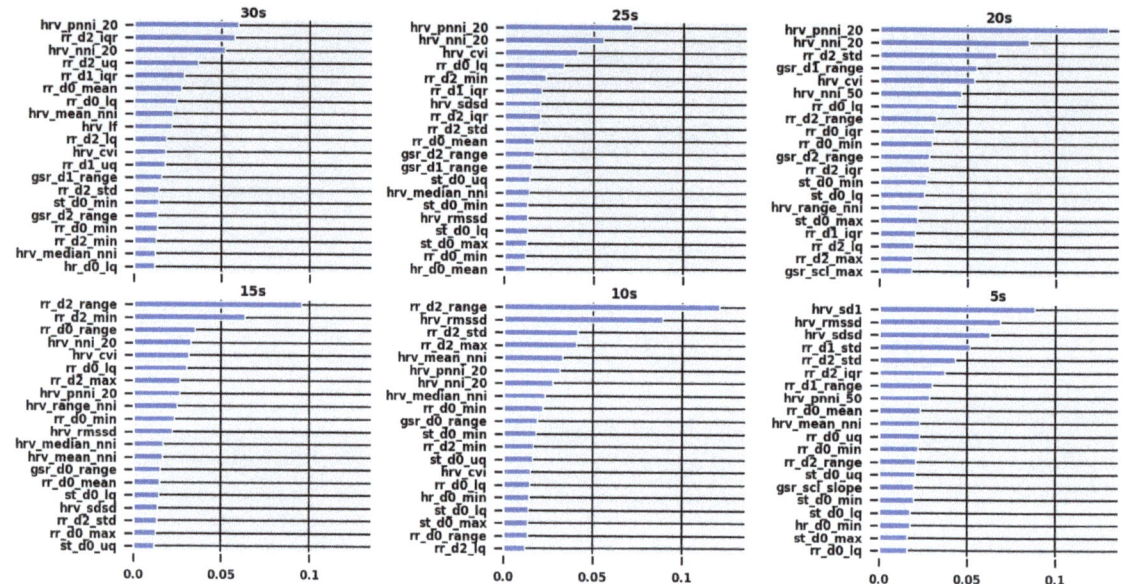

Figure 4. Importance of the top-20 features for each window length.

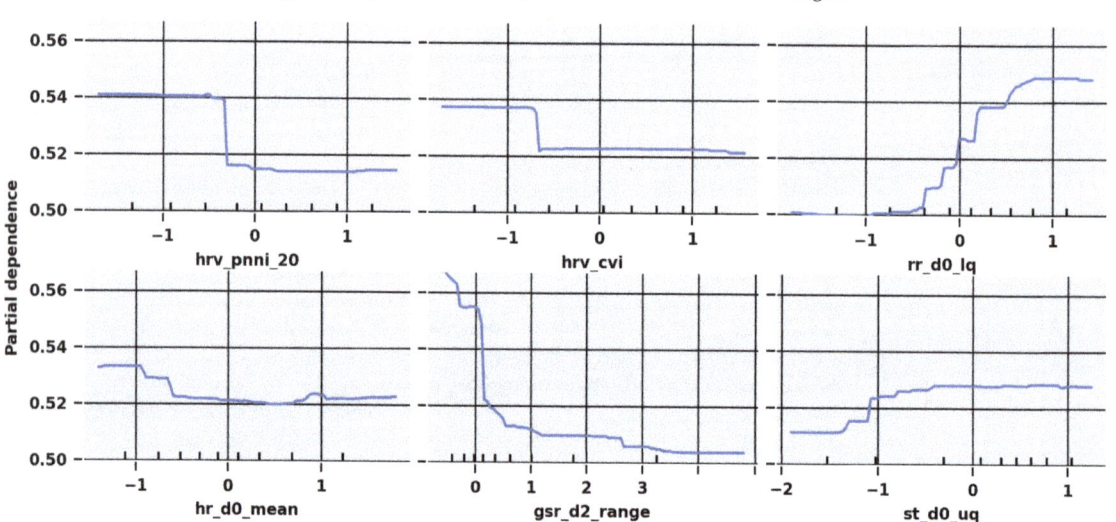

Figure 5. Partial dependency plots of selected features for the model with 25 s window length.

3.4. Individual Differences

In an ablation study, the performance of the XGB model was inspected with default hyperparameters without Bayesian optimization. The results for different window lengths and individuals are shown in Figure 6, with the means and confidence intervals computed across individuals in the train- and test-splits that the dataset came with. Overall, the mean accuracy increased as the window length increased until a window length of 25 s, and slightly decreased for the longest window length. The mean accuracy for the test-split was systematically higher than for the training-split, but the training accuracy was still within the confidence interval of the testing accuracy. However, the confidence interval of the mean test-split accuracy was obtained by bootstrapping five observations, so it is likely to be somewhat biased.

On an individual level, the accuracy between window lengths varied, but most users' individual accuracy (i.e., test accuracy when that individual was in the test fold during the LOSO validation) was at its maximum with a 25 s window length.

The individual variation in feature values is shown in Figure 7, which displays the boxplots of differences of the mean feature values between the cognitive load and resting state observed for each individual. The features selected to display were the same features as in Figure 5. The figure shows that usually the heart rate variability was higher (and, consequently, the RR was lower) while resting than when in the cognitive load state, the HR varied between individuals but was mostly higher in the resting state, the range of the second derivative of the GSR was higher in the resting state, and the skin temperature was lower in the resting state. In addition, the distribution of each variable contained positive and negative values: the physiological response to a cognitive load between individuals did not differ only in magnitude but also in the direction of the different responses.

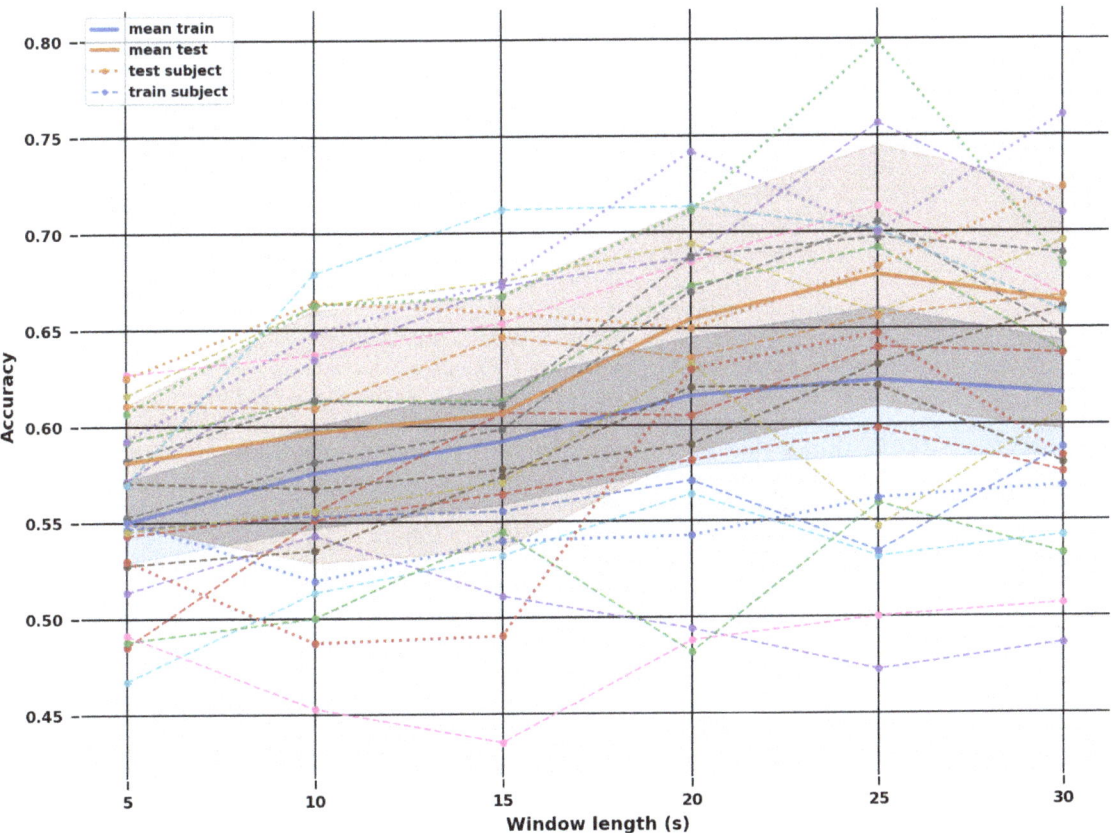

Figure 6. Classification accuracy observed with the dataset's default train-test split using the XGB model with default hyperparameters (ablating Bayesian optimization). Solid lines denote mean accuracy with confidence regions obtained by bootstrapping around them. Dashed lines depict the validation (training-split) and dotted lines the test (test-split) accuracy of a single subject.

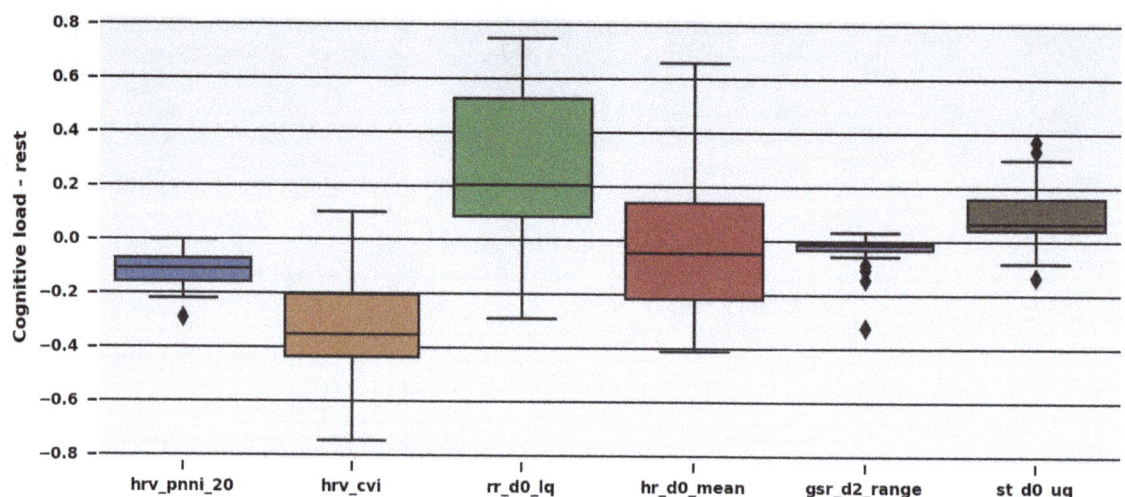

Figure 7. Difference of selected mean feature values between cognitive load and rest sessions across each individual, computed with 25 s window length. Positive values denote that the feature value was higher in cognitive load. Differences of rr_d0_lq and hrv_pnni_20 were scaled by dividing by 100, and hr_d0_mean by dividing by 10 to show the distribution of each feature better.

4. Discussion

The objective in this study was to compare the cognitive load detection performance at different ultra-short window lengths. Six window lengths of less than or equal to 30 s in duration were analyzed using a personalized approach with the XGBoost classifier and Bayesian hyperparameter optimization.

In terms of the overall classification accuracy, shorter windows showed lower performance and the best performance was found for windows of 25 s and 30 s with a statistically insignificant difference between the two. However, even if the differences between other window options were statistically significant, they were modest in absolute terms (lowest accuracy 60% vs. highest accuracy 67.6%). There were also large individual differences and person-specific factors which affected which samples were correctly and incorrectly classified. The individual accuracies ranged from 51% to 80% at a 25 s window length and the individual-specific optimal window length varied between 10 s, 20 s, 25 s, and 30 s.

Although earlier studies that have conducted experiments with different window lengths have not tested for statistical significance and they have mainly used window lengths above 30 s, the overall impression has been similar: longer windows tend to provide better performance. In [32], the best performance was found with a 120 s window and with one exception (15 vs. 30 s) the performance increased as the window length increased. The differences between the window lengths were small in [24], but still, longer windows performed better with nearly all of the tested classifiers.

Compared to the related work mentioned in Table 1, the classification accuracy in this study was rather low at each window length. Still, the highest accuracy (67.6%) at 25 s was almost the same as in the original paper [14] using the same dataset with a 30 s window (68.2%) and higher than in [45] (63.3%) and [46] (62%) using a subset of the same dataset and a 30 s window. The low performance is likely related to large individual differences and the tasks used in the dataset to elicit the cognitive load, which are discussed below.

The six elementary cognitive tasks (ECT) were selected based on [47], where they identified three relevant cognitive capabilities in the ubiquitous computing domain: flexibility of closure (HP), speed of closure (GC) and perceptual speed (FA, NC, PT, SX). However, the ECT refers to any range of basic tasks that require only a small number of mental processes and they have been originally designed to demonstrate individual differences between more than two participant groups (e.g., patients vs. healthy controls) [47]. Therefore, the cognitive load of these six tasks may have been mild compared to the N-back, the working memory task. Table 8 shows the task-wise accuracy for each of the tasks and despite the fact that there were less samples from N-back tasks than the other tasks, they were relatively well-recognized as a cognitive load.

In relation to real-life applications, however, eliciting a relatively mild cognitive load offers a more realistic situation. In real-life, extreme reactions (deep relaxation or high cognitive load) tend to occur rarely and reactions are milder than in laboratory protocols designed to elicit a high cognitive load or stress. All in all, the varying task difficulty between the seven different tasks, the three different task difficulty levels (low, moderate, high) used in the study, as well as the individual differences in cognitive performance and physiological reactions may have affected the varying classification results.

Table 8. The proportion of samples correctly classified as cognitive load during each cognitive task, and the number of windows for each task at a window length of 25 s.

	FA	GC	HP	NC	PT	SX	n2	n3
Proportion	0.723	0.469	0.845	0.750	0.638	0.747	0.839	0.730
Windows	382	207	336	388	210	245	168	174

Radüntz et al. [48] suggested that biomarkers, especially heart rate features, exhibit themselves on different timescales in cognitively demanding tasks. They found that the heart rate responded earlier to workload changes than frequency domain HRV parameters. This is in line with the findings in this paper that the most important variables for detecting a cognitive load were statistics of the RR intervals and HRV features from the time- and non-linear domain. Frequency domain variables were among top-20 only at a window length of 30 s, which may indicate that frequency domain measures respond slower than HR-related features from other domains. A similar notion was given in [22], who report that ultra-short frequency domain norms are from 20 to 180 s, and generally windows of 60 s and up to 24 h should be used.

Thus, the tasks were short enough that not all features could respond before the state changed again, which may have affected the feature importances and the direction where the features changed during the tasks. As evidenced in Figure 7, the direction of change between features and individuals varied, and e.g., the HRV was lower in the resting state than in the cognitive load state for some participants. Again, this may be a symptom of the tasks producing a mild cognitive load, but also in the way the resting and cognitive load states were defined. In this study, the states were defined as in the original paper, and the resting state was a combination of all resting periods before and between the cognitive tasks. However, the rest sessions located between the tasks are not similar to the resting state measured as a baseline before the tasks, or at the very end of the measurement protocol. Because people do not recover instantaneously, physiological reactions caused by cognitive tasks are still ongoing when the rest session begins, which likely affected the classification performance. A significantly higher classification performance was found, e.g., in [20], where the resting condition represented a baseline measurement conducted at the beginning and the end of the measurement protocol. However, these kinds of baseline rest periods have not been recorded in this dataset.

An analysis of confusion matrices revealed that errors made by classifying cognitive load periods as resting was quite stable for all the window lengths (between a minimum of 26.2% at 5 s and a maximum of 28.9% at 15 s) whereas the number of errors made by classifying resting as a cognitive load increased as the window length decreased (38.3% at 25 s and increasing to 53.8% at 5 s). Therefore, it seems that the classifier made the most errors when the person was recovering from a cognitive load and that the effect of the used division for cognitive load and resting state may have been especially strong for the shorter window lengths. Analyzing the dynamics of state detection in short windows might reveal more reasons why shorter windows had inferior performance, however, it is beyond scope of this text and is left for future work.

In this study, overlapping was used to utilize the available data more efficiently. Since each task lasted for as long as it took for the subject to complete it, the task length varied between tasks and individuals: easier tasks were completed faster and some subjects were quicker than others. All in all, approximately 34% of the tasks were completed in less than 60 s (needed to have two windows at a 30 s window length without an overlap), 16% in less than 45 s (needed to have two windows at 30 s window length with a 50% overlap), and 23% were completed in less than 50 s (needed to have two windows at a 25 s window length without an overlap) and 10% in less than 37.5 s (needed to have two windows at a 25 s window length with a 50% overlap). So, especially for the longer window lengths, overlapping increased the amount of available data and prevented disregarding data either from the beginning or the end of each task. Although overlapping is often employed in feature extraction (see Table 1) and a 50% overlap is commonly used in signal processing for spectral density estimation [49], its effect on the classification performance in cognitive state detection [9] and human activity recognition [50] has been found to be insignificant. However, overlapping can update the output incrementally and more efficiently than fixed windows [9] and thus it has potential value for future real-time systems with continuous state estimation.

The focus in this study was on cognitive load detection, which is methodologically closely related to affect, or emotion, recognition, where rather long windows are also often employed. Since an affective state tends to last for a very short time [21], shorter feature windows also for affect recognition should be investigated in future studies.

5. Conclusions and Future Work

Cognitive load assessment could serve multiple applications, e.g., in human–computer interaction to recognize and adapt to human overload issues. The future direction in assessing cognitive load is in real-time analysis and detecting the state in a streaming mode. In this study, a step towards more timely, real-time cognitive load detection was taken by analyzing the effect that ultra-short window lengths (30 s or less) have on detection performance.

The results on this dataset showed that longer windows perform better with statistically significant differences. The best performance of 67.6% was observed at a 25 s window length and the accuracy decreased to 60.0% at a 5 s window length. The optimal window length varied on an individual level, and whereas longer windows performed better on average, shorter windows were better for some individuals. Compared to earlier works using longer windows, the classification accuracies obtained were low, but the accuracy on the longer windows tested was similar or higher to those obtained earlier with the same dataset in [14,45,46] using a 30 s window.

The tasks used in the dataset produced rather mild cognitive load, which is closer to real-life circumstances but more difficult to detect than a higher load. Moreover, shorter windows contain less data, and some physiological features could not react on time to changes in the cognitive load and thus were not useful for state detection. R-to-R interval statistics as well as time- and non-linear domain HRV features had the fastest response to changes in cognitive load, followed by GSR statistics and skin temperature.

Short windows allow predicting the state more often, and so they may be more desirable in applications where more timely state detection is needed. However, shorter windows contain less data and physiological events, and so it is more difficult to correctly detect the state with shorter windows than it is with longer windows. Thus, the timeliness will be achieved on the expense of model accuracy as the results of this study demonstrate. The performance on a 5 s window was 7.6% behind of the performance on a 25 s window, despite that it contained five times less data. Although the performance found on this dataset was rather limited, this motivates future studies for real-time, even streaming, cognitive load detection.

Future studies towards this goal would benefit from a larger database to account for individual differences more effectively, to analyze the effects of window overlapping in terms of classification performance and continuous state detection, and to be able to use a larger set of different window lengths. Additionally, the analysis of the effect of short windows could be extended to other state detection tasks within affect recognition, to address similar issues in a broader context.

Supplementary Materials: The following are available online at https://www.mdpi.com/2079-9292/10/5/613/s1. The following supplementary files are available: posterior_distributions. pdf (figures similar to Figure 3 for all XGB hyperparameters), bayes_opt_results.csv (data generated containing information on each completed iteration of Bayesian optimization), and individual_accuracies.csv (subject-wise accuracies for each window length). The source code is available on Github at https://github.com/jatervon/ultra-short-cognitive-load-detection.

Author Contributions: Conceptualization, J.T., K.P. and J.M.; methodology, J.T., K.P. and J.M.; software, J.T.; validation, J.T., K.P. and J.M.; formal analysis, J.T.; investigation, J.T. and K.P.; resources, J.T., K.P. and J.M.; data curation, J.T.; writing—original draft preparation, J.T. and K.P.; writing—review and editing, J.M.; visualization, J.T.; supervision, J.M.; project administration, J.M.; funding acquisition, K.P. and J.M. All authors have read and agreed to the published version of the manuscript.

Funding: This research was funded by Academy of Finland grant number 334092.

Institutional Review Board Statement: Ethical review and approval were waived for this study, since no data was collected specifically for this study and a publicly available dataset was used instead.

Informed Consent Statement: Patient consent was waived for this study due to use of publicly available dataset.

Data Availability Statement: Publicly available dataset was analyzed in this study. Information on dataset with access link is available from Ref. [14].

Acknowledgments: The authors would like to thank the persons involved in collecting the dataset used and making it available as open data.

Conflicts of Interest: The authors declare no conflict of interest.

References

1. Krause, A.J.; Simon, E.B.; Mander, B.A.; Greer, S.M.; Saletin, J.M.; Goldstein-Piekarski, A.N.; Walker, M.P. The sleep-deprived human brain. *Nat. Rev. Neurosci.* **2017**, *18*, 404–418. [CrossRef] [PubMed]
2. Petruo, V.A.; Mückschel, M.; Beste, C. On the role of the prefrontal cortex in fatigue effects on cognitive flexibility—A system neurophysiological approach. *Sci. Rep.* **2018**, *8*, 1–13. [CrossRef]
3. Shields, G.S.; Sazma, M.A.; Yonelinas, A.P. The effects of acute stress on core executive functions: A meta-analysis and comparison with cortisol. *Neurosci. Biobehav. Rev.* **2016**, *68*, 651–668. [CrossRef]
4. Arnsten, A.F.T. Stress signalling pathways that impair prefrontal cortex structure and function. *Nat. Rev. Neurosci.* **2009**, *10*, 410–422. [CrossRef]
5. Sörqvist, P.; Dahlström, Ö.; Karlsson, T.; Rönnberg, J. Concentration: The neural underpinnings of how cognitive load shields against distraction. *Front. Hum. Neurosci.* **2016**, *10*, 1–10. [CrossRef] [PubMed]
6. Young, M.S.; Brookhuis, K.A.; Wickens, C.D.; Hancock, P.A. State of science: Mental workload in ergonomics. *Ergonomics* **2015**, *58*, 1–17. [CrossRef]
7. Won, E.; Kim, Y.K. Stress, the Autonomic Nervous System, and the Immune-kynurenine Pathway in the Etiology of Depression. *Curr. Neuropharmacol.* **2016**, *14*, 665–673. [CrossRef] [PubMed]
8. Kreibig, S.D. Autonomic nervous system activity in emotion: A review. *Biol. Psychol.* **2010**, *84*, 394–421. [CrossRef] [PubMed]
9. Anusha, A.S.; Jose, J.; Preejith, S.P.; Jayaraj, J.; Mohanasankar, S. Physiological signal based work stress detection using unobtrusive sensors. *Biomed. Phys. Eng. Express* **2018**, *4*. [CrossRef]
10. Vinkers, C.H.; Penning, R.; Hellhammer, J.; Verster, J.C.; Klaessens, J.H.G.M.; Olivier, B.; Kalkman, C.J. The effect of stress on core and peripheral body temperature in humans. *Stress* **2013**, *16*, 520–530. [CrossRef]
11. Larmuseau, C.; Cornelis, J.; Lancieri, L.; Desmet, P.; Depaepe, F. Multimodal learning analytics to investigate cognitive load during online problem solving. *Br. J. Educ. Technol.* **2020**, *51*, 1548–1562. [CrossRef]
12. Kistler, A.; Mariauzouls, C.; von Berlepsch, K. Fingertip temperature as an indicator for sympathetic responses. *Int. J. Psychophysiol.* **1998**, *29*, 35–41. [CrossRef]

13. Smets, E.; De Raedt, W.; Van Hoof, C. Into the Wild: The Challenges of Physiological Stress Detection in Laboratory and Ambulatory Settings. *IEEE J. Biomed. Health Inform.* **2019**, *23*, 463–473. [CrossRef] [PubMed]
14. Gjoreski, M.; Kolenik, T.; Knez, T.; Luštrek, M.; Gams, M.; Gjoreski, H.; Pejović, V. Datasets for cognitive load inference using wearable sensors and psychological traits. *Appl. Sci.* **2020**, *10*, 3843. [CrossRef]
15. Hidalgo-Muñoz, A.R.; Béquet, A.J.; Astier-Juvenon, M.; Pépin, G.; Fort, A.; Jallais, C.; Tattegrain, H.; Gabaude, C. Respiration and Heart Rate Modulation Due to Competing Cognitive Tasks While Driving. *Front. Hum. Neurosci.* **2019**, *12*, 1–8. [CrossRef]
16. Visnovcova, Z.; Mestanik, M.; Gala, M.; Mestanikova, A.; Tonhajzerova, I. The complexity of electrodermal activity is altered in mental cognitive stressors. *Comput. Biol. Med.* **2016**, *79*, 123–129. [CrossRef] [PubMed]
17. Castaldo, R.; Melillo, P.; Bracale, U.; Caserta, M.; Triassi, M.; Pecchia, L. Acute mental stress assessment via short term HRV analysis in healthy adults: A systematic review with meta-analysis. *Biomed. Signal Process. Control* **2015**, *18*, 370–377. [CrossRef]
18. Dehais, F.; Causse, M.; Vachon, F.; Tremblay, S. Cognitive conflict in human–automation interactions: A psychophysiological study. *Appl. Ergon.* **2012**, *43*, 588–595. [CrossRef]
19. Paprocki, R.; Lenskiy, A. What does eye-blink rate variability dynamics tell us about cognitive performance? *Front. Hum. Neurosci.* **2017**, *11*. [CrossRef]
20. Pettersson, K.; Tervonen, J.; Närväinen, J.; Henttonen, P.; Määttänen, I.; Mäntyjärvi, J. Selecting Feature Sets and Comparing Classification Methods for Cognitive State Estimation. In Proceedings of the 2020 IEEE 20th International Conference on Bioinformatics and Bioengineering (BIBE), Cincinnati, OH, USA, 26–28 October 2020; pp. 683–690. [CrossRef]
21. Schmidt, P.; Reiss, A.; Dürichen, R.; Laerhoven, K.V. Wearable-based affect recognition—A review. *Sensors* **2019**, *19*, 4079. [CrossRef] [PubMed]
22. Shaffer, F.; Ginsberg, J.P. An Overview of Heart Rate Variability Metrics and Norms. *Front. Public Health* **2017**, *5*, 258. [CrossRef]
23. Marshall, S.P. Identifying cognitive state from eye metrics. *Aviat. Space Environ. Med.* **2007**, *78*, B165–B175.
24. Gjoreski, M.; Luštrek, M.; Gams, M.; Gjoreski, H. Monitoring stress with a wrist device using context. *J. Biomed. Inform.* **2017**, *73*, 159–170. [CrossRef]
25. Smets, E.; Casale, P.; Großekathöfer, U.; Lamichhane, B.; De Raedt, W.; Bogaerts, K.; Van Diest, I.; Van Hoof, C. Comparison of machine learning techniques for psychophysiological stress detection. In *Pervasive Computing Paradigms for Mental Health. MindCare 2015, Communications in Computer and Information Science*; Springer: Cham, Switzerland, 2015; Volume 604, pp. 13–22. [CrossRef]
26. Castaldo, R.; Montesinos, L.; Melillo, P.; James, C.; Pecchia, L. Ultra-short term HRV features as surrogates of short term HRV: A case study on mental stress detection in real life. *BMC Med. Inform. Decis. Mak.* **2019**, *19*, 1–13. [CrossRef] [PubMed]
27. Huysmans, D.; Smets, E.; De Raedt, W.; Van Hoof, C.; Bogaerts, K.; Van Diest, I.; Helic, D. Unsupervised learning for mental stress detection exploration of self-organizing maps. In Proceedings of the BIOSIGNALS 2018—11th International Conference on Bio-Inspired Systems and Signal Processing, Part of 11th International Joint Conference on Biomedical Engineering Systems and Technologies (BIOSTEC 2018), Madeira, Portugal, 19–21 January 2018; pp. 26–35. [CrossRef]
28. Healey, J.; Nachman, L.; Subramanian, S.; Shahabdeen, J.; Morris, M. Out of the Lab and into the Fray: Towards Modeling Emotion in Everyday Life. In *Pervasive Computing*; Floréen, P., Krüger, A., Spasojevic, M., Eds.; Springer: Berlin/Heidelberg, Germany, 2010; pp. 156–173.
29. Marín-Morales, J.; Higuera-Trujillo, J.L.; Greco, A.; Guixeres, J.; Llinares, C.; Scilingo, E.P.; Alcañiz, M.; Valenza, G. Affective computing ual reality: Emotion recognition from brain and heartbeat dynamics using wearable sensors. *Sci. Rep.* **2018**, *8*, 1–15. [CrossRef] [PubMed]
30. Guo, H.W.; Huang, Y.S.; Lin, C.H.; Chien, J.C.; Haraikawa, K.; Shieh, J.S. Heart Rate Variability Signal Features for Emotion Recognition by Using Principal Component Analysis and Support Vectors Machine. In Proceedings of the 2016 IEEE 16th International Conference on Bioinformatics and Bioengineering (BIBE 2016), Taichung, Taiwan, 31 October–2 November 2016; pp. 274–277. [CrossRef]
31. Stikic, M.; Johnson, R.R.; Tan, V.; Berka, C. EEG-based classification of positive and negative affective states. *Brain-Comput. Interfaces* **2014**, *1*, 99–112. [CrossRef]
32. Siirtola, P. Continuous stress detection using the sensors of commercial smartwatch. In *Proceedings of the UbiComp/ISWC 2019—Adjunct Proceedings of the 2019 ACM International Joint Conference on Pervasive and Ubiquitous Computing and Proceedings of the 2019 ACM International Symposium on Wearable Computers*; ACM: London, UK, 2019; pp. 1198–1201. [CrossRef]
33. Schmidt, P.; Reiss, A.; Duerichen, R.; Marberger, C.; Van Laerhoven, K. Introducing WESAD, a multimodal dataset for wearable stress and affect detection. In Proceedings of the 2018 on International Conference on Multimodal Interaction (ICMI '18), Boulder, CO, USA, 16–20 October 2018; ACM Press: New York, NY, USA, 2018; pp. 400–408. [CrossRef]
34. Kroupi, E.; Vesin, J.M.; Ebrahimi, T. Subject-Independent Odor Pleasantness Classification Using Brain and Peripheral Signals. *IEEE Trans. Affect. Comput.* **2016**, *7*, 422–434. [CrossRef]
35. Bota, P.J.; Wang, C.; Fred, A.L.N.; Placido Da Silva, H. A Review, Current Challenges, and Future Possibilities on Emotion Recognition Using Machine Learning and Physiological Signals. *IEEE Access* **2019**, *7*, 140990–141020. [CrossRef]
36. Jeppesen, J.; Beniczky, S.; Johansen, P.; Sidenius, P.; Fuglsang-Frederiksen, A. Using Lorenz plot and Cardiac Sympathetic Index of heart rate variability for detecting seizures for patients with epilepsy. In Proceedings of the 2014 36th Annual International Conference of the IEEE Engineering in Medicine and Biology Society (EMBC 2014), Chicago, IL, USA, 26–30 August 2014; pp. 4563–4566. [CrossRef]

37. Tervonen, J.; Puttonen, S.; Sillanpää, M.J.; Hopsu, L.; Homorodi, Z.; Keränen, J.; Pajukanta, J.; Tolonen, A.; Lämsä, A.; Mäntyjärvi, J. Personalized mental stress detection with self-organizing map: From laboratory to the field. *Comput. Biol. Med.* **2020**, *124*, 103935. [CrossRef]
38. Chen, T.; Guestrin, C. XGBoost: A scalable tree boosting system. In Proceedings of the 22nd ACM SIGKDD International Conference on Knowledge Discovery and Data Mining, San Francisco, CA, USA, 13–17 August 2016; ACM: New York, NY, USA, 2016; Volume 19, pp. 785–794. [CrossRef]
39. Fernández-Delgado, M.; Cernadas, E.; Barro, S.; Amorim, D. Do we need hundreds of classifiers to solve real world classification problems? *J. Mach. Learn. Res.* **2014**, *15*, 3133–3181. [CrossRef]
40. Frazier, P.I. A Tutorial on Bayesian Optimization. *arXiv* **2018**, arXiv:1807.02811.
41. Pedregosa, F.; Varoquaux, G.; Gramfort, A.; Michel, V.; Thirion, B.; Grisel, O.; Blondel, M.; Prettenhofer, P.; Weiss, R.; Dubourg, V. Scikit-learn: Machine Learning in Python. *J. Mach. Learn. Res.* **2011**, *12*, 2825–2830.
42. Champseix, R. Heart Rate Variability Analysis. 2018. Available online: https://github.com/Aura-healthcare/hrvanalysis (accessed on 18 January 2021).
43. Makowski, D.; Pham, T.; Lau, Z.J.; Brammer, J.C.; Lespinasse, F.; Pham, H.; Schölzel, C.; Chen, S.H.A. NeuroKit2: A Python Toolbox for Neurophysiological Signal Processing. *Behav. Res. Methods* **2020**. [CrossRef]
44. Bergstra, J.; Yamins, D.L.K.; Cox, D.D. Making a Science of Model Search: Hyperparameter Optimization in Hundreds of Dimensions for Vision Architectures. In Proceedings of the 30th International Conference on Machine Learning (ICML'13), Atlanta, GA, USA, 16–21 June 2013; Volume 28, pp. 115–123.
45. Li, X.; De Cock, M. Cognitive load detection from wrist-band sensors. In *UbiComp/ISWC 2020 Adjunct—Proceedings of the 2020 ACM International Joint Conference on Pervasive and Ubiquitous Computing and Proceedings of the 2020 ACM International Symposium on Wearable Computers*; ACM: Cancun, Mexico, 2020; pp. 456–461. [CrossRef]
46. Salfinger, A. Deep learning for cognitive load monitoring: A comparative evaluation. In *UbiComp/ISWC 2020 Adjunct—Proceedings of the 2020 ACM International Joint Conference on Pervasive and Ubiquitous Computing and Proceedings of the 2020 ACM International Symposium on Wearable Computers*; ACM: Cancun, Mexico, 2020; pp. 462–467. [CrossRef]
47. Haapalainen, E.; Kim, S.; Forlizzi, J.F.; Dey, A.K. Psycho-physiological measures for assessing cognitive load. In *Proceedings of the 2010 ACM Conference on Ubiquitous Computing*; ACM: Copenhagen, Denmark, 2010; pp. 301–310. [CrossRef]
48. Radüntz, T.; Mühlhausen, T.; Freyer, M.; Fürstenau, N.; Meffert, B. Cardiovascular Biomarkers' Inherent Timescales in Mental Workload Assessment During Simulated Air Traffic Control Tasks. *Appl. Psychophysiol. Biofeedback* **2020**. [CrossRef] [PubMed]
49. Heinzel, G.; Rüdiger, A.; Schilling, R. Spectrum and Spectral Density Estimation by the DISCRETE Fourier Transform (DFT), Including a Comprehensive List of Window Functions and Some New at-Top Windows. (unpublished). **2002**, 1–84. Available online: https://holometer.fnal.gov/GH_FFT.pdf (accessed on 18 January 2021)
50. Dehghani, A.; Sarbishei, O.; Glatard, T.; Shihab, E. A Quantitative Comparison of Overlapping and Non-Overlapping Sliding Windows for Human Activity Recognition Using Inertial Sensors. *Sensors* **2019**, *19*, 5026. [CrossRef]

Article

A One-Dimensional Non-Intrusive and Privacy-Preserving Identification System for Households

Tomaž Kompara [1,2,*], Janez Perš [3,*], David Susič [1,*] and Matjaž Gams [1,2,*]

1. Department of Intelligent Systems, Jožef Stefan Institute, Jamova Cesta 39, 1000 Ljubljana, Slovenia
2. Jožef Stefan International Postgraduate School, Jamova Cesta 39, 1000 Ljubljana, Slovenia
3. Faculty of Electrical Engineering, University of Ljubljana, Tržaška Cesta 25, 1000 Ljubljana, Slovenia
* Correspondence: tomaz.kompara@ijs.si (T.K.); janez.pers@fe.uni-lj.si (J.P.); david.susic@ijs.si (D.S.); matjaz.gams@ijs.si (M.G.)

Citation: Kompara, T.; Perš, J.; Susič, D.; Gams, M. A One-Dimensional Non-Intrusive and Privacy-Preserving Identification System for Households. *Electronics* **2021**, *10*, 559. https://doi.org/10.3390/electronics 10050559

Academic Editor: Daniele Riboni

Received: 17 January 2021
Accepted: 23 February 2021
Published: 27 February 2021

Publisher's Note: MDPI stays neutral with regard to jurisdictional claims in published maps and institutional affiliations.

Copyright: © 2021 by the authors. Licensee MDPI, Basel, Switzerland. This article is an open access article distributed under the terms and conditions of the Creative Commons Attribution (CC BY) license (https:// creativecommons.org/licenses/by/ 4.0/).

Abstract: In many ambient-intelligence applications, including intelligent homes and cities, awareness of an inhabitant's presence and identity is of great importance. Such an identification system should be non-intrusive and therefore seamless for the user, especially if our goal is ubiquitous and pervasive surveillance. However, due to privacy concerns and regulatory restrictions, such a system should also strive to preserve the user's privacy as much as possible. In this paper, a novel identification system is presented based on a network of laser sensors, each attached on top of the room entry. Its sensor modality, a one-dimensional depth sensor, was chosen with privacy in mind. Each sensor is mounted on the top of a doorway, facing towards the entrance, at an angle. This position allows acquiring the user's body shape while the user is crossing the doorway, and the classification is performed by classical machine learning methods. The system is non-intrusive, non-intrusive and preserves privacy—it omits specific user-sensitive information such as activity, facial expression or clothing. No video or audio data are required. The feasibility of such a system was tested on a nearly 4000-person, publicly available database of anthropometric measurements to analyze the relationships among accuracy, measured data and number of residents, while the evaluation of the system was conducted in a real-world scenario on 18 subjects. The evaluation was performed on a closed dataset with a 10-fold cross validation and showed 98.4% accuracy for all subjects. The accuracy for groups of five subjects averaged 99.1%. These results indicate that a network of one-dimensional depth sensors is suitable for the identification task with purposes such as surveillance and intelligent ambience.

Keywords: one-dimensional depth sensor; biometrics; identification; machine learning

1. Introduction

For many years, Artificial Intelligence (AI) has played a central role in techniques that improve system performance in various areas, especially when Machine Learning (ML) has been used. The immense growth of data due to the Internet and Internet of Things, as well as the increase in computing power, has led to a great increase in the benefits of AI in the last decade [1], as there are many data to learn from, and the amount and speed of data exceeds human capabilities. Therefore, AI has a significant impact on people's daily lives nowadays. Ambient Intelligence (AmI) enhances people's everyday lives by sensing their presence and responding to their actions [2,3]. To provide this service without interfering with the users' activities, the sensing must be non-intrusive. This means that the users perform their activities in exactly the same way as if the sensors did not exist. Furthermore, the sensing should preserve the users' privacy as far as possible, so that such systems can be used in private environments such as smart homes [4]. In addition, sensors should be robust, low cost and highly accurate to be used in real life situations. These are the main requirements for AmI sensors that need to be met in order to introduce them into everyday life.

The key information that allows ambience to adapt to the individual user is the number of people who are present and their identities [5]. This information is helpful in order to put additional sensor information into the proper context. For example, a user comes home and turns a thermostat to 20 °C. After a while, his wife comes home and sets the thermostat to 22 °C, since this is the temperature she is comfortable at. If the users' identities are provided, the AmI system can regulate the temperature of the room according to the preferences of the users who are present, without any further input. Similar problems affect cooling, ventilation, assistance for the elderly, security systems, etc. In the same way, non-essential functions, such as the choice of music, movies and even commercials, can be controlled based on the identity of the people who are present. Furthermore, the history of the inhabitants' presence can be used to predict their next action and adjust the ambient environment in advance (controlling robot cleaners, domestic hot-water preparation and powering standby devices) [6]. Therefore, detecting the number and identities of the people who are present is one of the preconditions for successful AmI applications. It should be noted that the identification task for setting room preferences is quite different in nature compared to the identification when entering a smart home. The inability to classify correctly an entry in a home can lead to severe consequences, whereas the smart home taking care of room residents can always set the preferences to default values if the person entering a room is not classified with sufficient probability.

Laser-based technology has made remarkable progress in recent decades. It covers a wide range of fields, such as medical sciences, space sciences and military technologies [7,8]. Laser sensors are capable of detecting, counting, imaging and scanning distances and proximity, making them ideal for numerous applications such as vehicle automation and guidance, traffic management, security and surveillance and warehouse management. They are therefore ideal when it comes to home sensor applications such as user silhouette detection and authentication.

In recent years, many non-intrusive identification systems were designed; however, most of them are not able to preserve privacy, while at the same time obtaining a high identification accuracy. This weakness was the motivation behind our research. For the sensor to be non-intrusive, no device has to be carried by the user for the identification to be successful, nor is any additional interaction required—a typical example of such needless interaction would be putting a finger on a fingerprint scanner. For practical purposes, this narrows the sensor selection down to measuring the biometrically relevant physical properties of people. In our case, the shape of the human body was selected as the main measure. To preserve privacy, a one-dimensional depth sensor mounted on the top of a doorway, looking downward at an angle, was used. In this way, the person's body shape is not captured with a single shot, but with multiple measurements during the whole doorway-crossing event. Effectively, the sensor follows the best practice for designing surveillance systems, as set forth by privacy regulators worldwide, known as Privacy by Design: the system should acquire only the essential data required to solve the problem it addresses. This means that such a system cannot provide more data than it needs, even in the event of a third-party intrusion; all the sensor obtains is a partial, one-sided and relatively low-resolution depth map of a person, and that is all that an attacker could possibly gain. If, for example, a live video feed were used to recognize people, this would mean potentially catastrophic privacy consequences in the event of an attack, especially if such devices are used in private environments. On the other hand, the classification accuracy should remain as high as possible even with such "blurred" data having in mind that in a private home only a few people are to be classified. Finally, we are not interested in exact indoor location to set the room ambient parameters. These are the main assumptions leading to our approach. In the following, formal definitions of being invasive, intrusive and privacy-preserving are introduced:

Definition 1 (Invasive). *The term invasive in this paper encompasses involving entry into the living body, e.g., by an incision or insertion of an instrument, or similar entry into the mental or cognitive state, including the implantation of changes in the well-being or emotions [9].*

In plain English, to be invasive is to have something introduced into the user's physical or mental state, either voluntarily or involuntarily. Physically, there does not have to be an actual physical insertion into the body, for example anything that applies pressure to any part of the user's physical state is considered invasive since it changes the physical state at the point of pressure. Psychologically, it is a bit similar and a bit different in that no physical influence is required, but the cognitive or psychological effect is similar—if something is introduced (or pushed) into the user's mental state or behavior by some device or system, it is considered invasive in the non-physical sense.

Definition 2 (Intrusive). *In this paper, the term intrusive refers to the disruption of user's normal behavior [10].*

By analogy to the definition of invasive, the term intrusive is physical, mental or both. In addition, the term intrusive is often relative and refers to the culturally accepted activities in a particular community. For example, if a person is to enter a home, a certain activity is required to pass a security test, such as unlocking a door with a door key. If such entry is generally accepted, it is considered non-intrusive. However, if a camera system is introduced that unlocks the door based on facial recognition when a user enters the department, this system is considered non-intrusive and the prior entry with the door key becomes intrusive as a user has to unlock the door compared to simply approaching the door when facial recognition is used. Another example of change would be the automatic unlocking of the door of some modern cars which happens when a user approaches with a key, which is obviously non-intrusive, and the current normal unlocking of the car door by pressing a button or using a car key now becomes intrusive.

This paper is about passing an unlocked interior door that is without doors, with opened doors or needs to be opened in some way. If a user passes the door as usual and no additional load is introduced by the AmI system, this is considered non-invasive and non-intrusive.

Definition 3 (Privacy-preserving). *The term privacy-preserving in this paper refers to the concept of security or harmlessness of user data when the data are or could be transmitted or communicated between different parties. The other party is not able to draw a potentially harmful conclusion from the data obtained [11].*

An example of non-privacy-preserving data are images taken with a camera. Even without actual security issues, users are usually uncomfortable with the knowledge that some device is taking accurate pictures of them. On the other hand, if the image captured by the camera is blurry enough that no one can see anything potentially harmful from it, it is considered privacy-preserving. This term is—as the previous two terms—in contradiction with detection accuracy. Nevertheless, there are systems that are both privacy-preserving and sufficiently accurate—for example, a left–right blurred image caused by a mechanical lens might allow correct height detection. In reality, systems of different types each establish their own relationship between these properties and accuracy, trying to accommodate for user needs and preferences.

Our approach is based on two hypothesis:

Hypothesis 1. *Due to the advancements of the laser devices and AI, the proposed laser-based system using AI methods will enable highly accurate identification of a small number of residents in a typical home.*

Hypothesis 2. *The introduced system will be non-invasive, non-intrusive and privacy-preserving.*

The process of data collection is described in detail to show that Hypothesis 2 is indeed satisfied, while the measurements of classification accuracy reveal relationships among various factors that affect identification accuracy to show conditions under which Hypothesis 1 is valid.

Since no such device exists in the mass market, the task is quite challenging and involves another parameter: the cost scheme. The goal is to use a device that should not exceed $100 as a general threshold.

In the remainder of the paper, we first present related work. After the related work, we describe the preliminary study conducted using the publicly available anthropometric measurements database. Next, we describe our system setup and its geometry. We continue with a description of the feature extraction process that describes the user's body shape. We then describe how the extracted features are used to determine direction and identity, as well as the evaluation process, its results and the comparison between theoretical estimates from a database and practical measurements. Finally, we present a discussion and conclusions with pros and cons.

2. Related Work

In the past decade, we have witnessed many attempts to develop identification systems that could support AmI applications on a ubiquitous scale and that could be easily integrated into the environment itself.

The two main application requirements are a high identification accuracy and non-intrusiveness. The first requirement arises from the need to correctly identify a person in order to properly personalize the environment. Every misidentification could lead to a user's discomfort, security risks and/or non-optimal energy use. The second requirement allows the user to maintain his/her way of living and interaction with the environment in exactly the same way as if the identification system did not exist. To meet both requirements, different sensors and techniques were explored in the scientific community. The most promising non-intrusive identification methods are the following:

- Pressure sensors are installed in the floor and used for measuring the location and the force of a foot. The user has to step onto the sensed area where the sensor is installed to be identified. It has been demonstrated that people can be identified according to the force profile of their footsteps [12,13]. Orr and Abowd were able to achieve 93% precision using 15 test people and a variety of footwear. Middleton et al., on the other hand, used an array of binary contact sensors and achieved 80% precision using the same number of test people.
- Doppler-shift sensors can determine an object's velocity based on a reflected wave's frequency shift caused by a moving object. According to Kalgaonkar and Raj [14], this sensor can be used to identify users based on their walking signatures when they walk straight towards the sensor. They obtained 90% accuracy for 30 test subjects in a laboratory environment, where only a single subject was observed at a time.
- Cameras are the most widely explored and used non-intrusive identification sensors. They are used to identify people with both face and gait recognition. Face-recognition methods much depend on lighting, facial expression, rotation, number of training examples and similar parameters [15]. Reported precision values vary widely between 37% and 98% [16]. Gait-recognition methods are mostly based on a person's silhouette dynamics. An accuracy of 87% was achieved by Tao et al. [17] for a single person walking on different surfaces. This number falls dramatically to 33% when the view angles, shoe types and surface types are varied. In some environments, people's activities and motion can be used to recognize identity, as in [18], where an 82% accuracy is observed. There were also some attempts to use extremely low temporal and spatial resolution cameras; however, the collected data can also be used for activity recognition, which may be undesirable [19]. Recent work in this field focuses mostly on much more difficult re-identification problem (e.g., [20,21]). Recent approaches in this field are based on deep learning (e.g., [22–24]).

- Scanning range-finders emit a signal and determine the distance of an object according to the time of flight, phase shift, or reflected angle. Such sensors identify people according to their body dimensions, which is similar to our approach. By using range-finders only the distance of a single point, a line or a full two-dimensional depth map can be obtained. On the basis of multiple single-point sensors, a 90% accuracy on three test subjects was achieved by Hnat et al. [25], whereas a 94% accuracy was obtained on eight test subjects by Kouno et al. [26] using full, two-dimensional depth images, obtained with the widely available Microsoft Kinect sensor.
- A radar-based system mounted at the top of the doorway that analyzes the signals reflected back from the environment to perform the identification is described in [27]. The identification accuracy for a group of an eight people was 66%. Nonetheless, this approach is proven to be good at people presence estimation [28].
- The thorough description of human monitoring system based on WiFi connectivity, in a fashion strictly compliant with the Internet of Things (IoT) paradigm, is given in [29]. In [30], a human identification system that leverages unique fine-grained gait patterns from existing WiFi-enabled IoT is proposed. The system achieves an average human identification accuracy of 91% from a group of 20 people.
- In [31], height based approach using a thermal camera is presented. The authors reported 92.9% accuracy for people with more than 2.5 cm difference in height tested on 21 subjects.
- In [32], a system is described that exploits pulsing signals from the photoplethysmography sensor in wrist-worn wearables for continuous user authentication. The system relies on the uniqueness of the human cardiac system and requires little training for successful deployment. In combination with the location recognition methods already mentioned, this system shows great potential as it enables seamless user authentication. However, it is not entirely non-intrusive, as the user must carry a wrist-worn wearable.

The authors would like to draw the reader's attention to some other systems that deal with privacy and authentication based data collection. Mobile crowd sensing is a sensing paradigm that uses sensor data from mobile devices carried by humans from a crowd of participants. Such data aggregation systems need to ensure the privacy of each participant [33]. A novel approach that combats this problem very successfully is presented in [34]. The mobile crowd-sensing system is not directly related to our work, as it aggregates data from all participants and does not focus on authenticating each participant. Privacy-preserving authentication in information retrieval from geographically based social networks is described in [35]. This method differs from ours in that it depends on knowing the geographic locations of each end system contained in the network, whereas our approach identifies the user only when it passes through the door.

It is our belief—and a strongly expressed view of privacy regulators, especially in the EU—that a high identification accuracy and non-intrusiveness are not the only requirements that should be considered. Privacy invasiveness should also be taken seriously, since the risk–benefit ratio is an important aspect, both when users are deciding whether or not to use the system and when privacy regulators are deciding whether to allow it for a particular purpose or not (as the [36] states: "Member States shall provide that personal data must be ... adequate, relevant and not excessive in relation to the purposes for which they are collected and/or further processed..."). Sensor solutions for AmI are especially sensitive in this aspect, as they mainly provide higher comfort to the users, and therefore the privacy threshold is particularly high. Privacy regulators are bound to be more conservative when evaluating such solutions, unlike the applications where security or people's lives are at a stake.

Although all identification systems can be—and usually are—designed to prevent third-party access, there is always a security risk from an adversary attack, which might expose privacy-sensitive data. One of the most promising approaches to promote privacy and data-protection compliance from the start is Privacy by Design [37,38]. A key guideline to reduce the privacy risk is through minimizing the amount of data, meaning only absolutely necessary information is acquired and used in the process, and only a bare minimum is stored.

From the identification system's perspective, the amount of acquired data should be adjusted according to the required identification accuracy and the purpose of use. For example, in high-security environments (e.g., banks and nuclear power plants), the identification accuracy should be as high as possible, regardless of the privacy invasiveness. On the other hand, in private environments, e.g. people's homes, where the identification system mostly controls entertainment electronics and other appliances, privacy has a higher priority than accuracy. Therefore, to make identification systems suitable for home use, the privacy issues need to be resolved. For example, cameras can be used not only for person identification at a doorway crossing, but for other purposes as well, such as recognizing users' activities, clothing, social behavior and interaction. According to the Privacy by Design approach, such an amount of acquired data is excessive for the identification task alone. Therefore, neither camera-based nor a two-dimensional depth-sensor-based approach is suitable for the identification task in private environments without further constraints on data acquisition.

On the other hand, unlike surveillance and security systems, many tasks of the intelligent environment do not require extreme positional accuracy. For example, tasks such as personalized adjustment of room temperature, lightning and (possibly) music volume do not require precise localization of people; this results in both less expensive solutions and greater privacy protection. Additionally, in home environments, the number of people is small and we assume they are cooperative, e.g., it is not in their interest to seek out faults in the system, but they may be highly sensitive to privacy issues. Our approach is intended for such cases.

Finally, the 1D laser scanner we use may be considered mature or even obsolete in terms of technology, but, again, this is not so if privacy concerns are a factor. In comparison to other sensors now widely used in home environment (e.g., Asus Xtion Pro and PrimeSense Carmine [39] provide 2D depth with accuracy about 4 mm in depth axis), our sensor can be seen simply as single scanline equivalent. It is important to understand that the latest methods using sensors such as Kinect v2 are able to infer very private information about people, such as facial expressions [40] and even comprehension [41]. Therefore, the moment such sensor is introduced into the home, it opens the door wide open to privacy violations.

The two approaches most similar to ours are found in [25,26]. Both rely on the subject's body shape for identification and both use depth sensors. However, the one in [25] only measures the depth at a single point, making it significantly less accurate, but, privacy-wise, it is very non-invasive. On the other hand, the one in [26] uses a Kinect sensor, obtaining a dense depth map. Although not used in [26], Kinect provides an RGB camera, which makes it a significant privacy risk, and a dense depth map has a significant field of view (therefore, enabling an adversary to observe additional activity in the area, which was not meant to be observed). In contrast to this, our approach provides high accuracy, but observes the environment only through a single, static, one-dimensional depth sensor and relies instead on the motion of the people across the doorway to acquire any useful information. Therefore, it cannot be used to observe the environment and the user's activity in any other way—except when a person moves through the doorway. Any other activity (e.g., a person standing and gesticulating) would result in a garbled depth map, unlike with 2D depth sensors or cameras.

3. Identifying People from Body Measurements

Related work shows that people can be identified to a certain degree even from small amount of information regarding their (apparent) shape. There are many approaches that use different sensor modalities, but they all rely on shape information:

- In some cases, researchers found that a single height measurement solves their problem; this could be done using scanning range-finders [25] or thermal camera [31].
- The next step in complexity is using a 2D sensor, e.g., camera, to obtain silhouette shape [20].
- A 2D sensor may be augmented with depth estimation (e.g., Kinect 3D sensor), which allows multiple anthropometric measurements from depth image data [21].
- Finally, ever more popular deep learning approaches actually learn which features they will use [24]. Algorithms are opaque, but we may assume that shape, whenever distinct enough for (re)identification, will be included at some point in the computation hierarchy.

Our approach assumes that the subjects can be identified using a limited number of body measurements of limited quality, as this intrinsically safeguards privacy. Before building the actual people-identification system, we examined the theoretical limits with regards to the accuracy of body-shape-based identification systems.

For this part of the study, we used the ANSUR database [42], which has 132 anthropometric measurements for 3982 subjects aged between 17 and 51. The measurements in the database were obtained manually with an accuracy in the millimeter range. However, the data provided by the depth sensor cannot achieve such high-level accuracy, especially in real-life environments. The main sources of noise for the measurement are different ways of dressing, different hair styles, body-weight variations, the objects being carried, etc. To simulate these influences on the identification accuracy, a uniform noise was added to the anthropometric measurements.

The relation among the peak-to-peak noise amplitude i, the number of people N and the identification accuracy E based on a single measure can be estimated using Equation (1). The measurement distribution p was obtained from the ANSUR database.

$$E(i) = \int_{-\infty}^{+\infty} 1 - \left(\int_{x-i/2}^{x+i/2} p(y) dy \right)^N dx \qquad (1)$$

However, the computational complexity increases significantly with a growing number of subjects and the number of measurement combinations, which prevented us from an in-depth analysis of the identification accuracy for larger sets of measurements. Therefore, we decided to perform a large-scale simulation experiment based on randomly picking two, three and more subjects from the ANSUR database and identifying them. We chose four measurements that can be reliably obtained by our sensor: height, height of the shoulders, shoulder width and head width, as part of the silhouette. Based on those four measurements, the identification accuracy was estimated in the presence of a varying amount of noise.

The accuracy was calculated based on one million randomly chosen groups of 2–10 subjects. The chosen amount of noise was added to their measurements. The result, i.e., the recognition accuracy, which depends on the number of subjects and the amount of noise, is shown in Figure 1. Each point on the grid on the surface in Figure 1 represents one million random trials. The nearest neighbor algorithm [43] was used to determine the identity of the noised data. If the classified identity corresponded to the ground truth from the database, it was considered as correct; otherwise, it was considered as a false identification. In addition to using all four chosen measurements, we performed tests on the subsets of those measurements. In Figure 1, the identification accuracy of all four measurements, the best-performing combination of two measurements and the best single measurement are presented. The single measurement results were additionally verified using Equation (1) and provided nearly identical results.

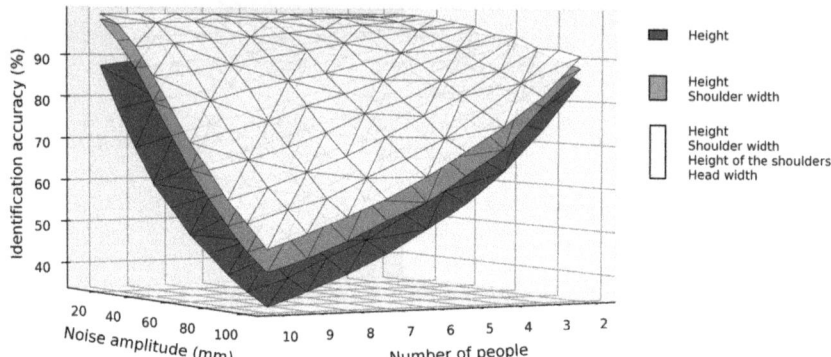

Figure 1. Theoretical identification accuracy according to the different measures, number of people and peak-to-peak noise amplitudes.

This analysis shows that combining more measurements increases the identification accuracy and can counteract the effects of noise, which significantly degrades the accuracy. Combining more measurements also allows us to identify subjects in larger groups, since the accuracy decreases with an increasing number of subjects as well. Note that this simulation was performed on four selected measurements that can be acquired with our sensor, while the developed one-dimensional depth sensor is capable of acquiring measurements of the whole body, i.e., not only shoulder width, for example, but also other widths, e.g., of the hips. On the other hand, Figure 1 indicates theoretic measurements with real-life obstacles such as dress or carried objects will decrease identification accuracy, which might make the identification useless in particular for larger groups of people. For a typical family, however, the number of family members is rather small, family members are usually of different heights and combining more measurements increases the overall accuracy, therefore Figure 1 provides an indication that our approach might be promising for home applications.

4. Sensor-Based Data Extraction

In this section, we describe the sensor device, the process of capturing data, identification and determination of the crossing direction from one-dimensional depth-sensor data, along with the sensor geometry and the overall system architecture. The transformation of the sensor data is applied to ensure data normalization, i.e., to ensure the compatibility of the acquired data between the sensors, regardless of the sensor characteristics and the mounting position. After the transformation procedure, the features are extracted and passed to the crossing-direction detection and the people-identification methods. While the basic approach presented here is not particularly novel, its understanding is important to properly describe the silhouette creation.

4.1. Sensor Geometry

The sensor is mounted on the top of the doorway facing down at an angle, as shown in Figure 2. Its mounting position allows the acquisition of one side of the user's body in a single doorway-crossing event. To ensure good-quality data, the sensor has to be set properly, as follows.

As shown in Figure 2, the sensor effectively projects a laser plane, which, by intersecting with the object of interest, forms a line. The plane's slope (α in Figure 2) depends on the sensor's mounting angle. Therefore, the angle α determines the size of the acquisition area. A smaller α results in a larger measuring area, and vice versa. A small measuring area is preferred, since only the people passing should be measured to preserve the privacy of other users, who may be present in a room, but are currently not passing through the doorway. However, with a smaller measuring are, a a higher sampling frequency is required to obtain sufficiently detailed data. For that reason, the sensor's mounting angle should be chosen according to the sensor's sampling frequency.

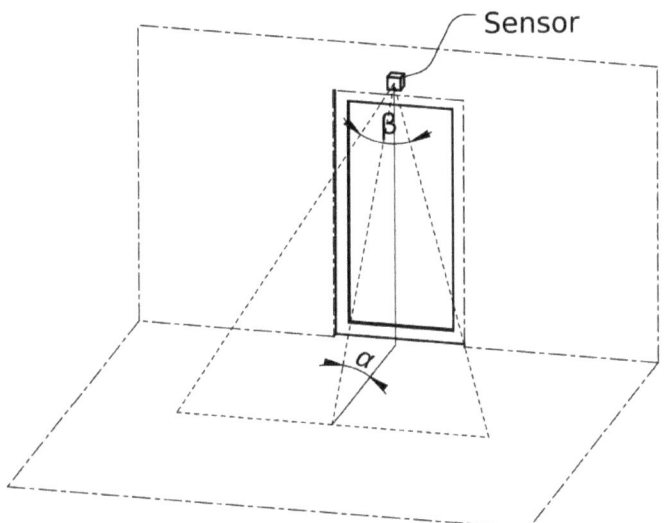

Figure 2. One-dimensional depth sensor mounted on the top of the doorway.

Other sensor properties that affect the quality of the measurements are the field of view (β in Figure 2), the angular resolution and the raw sensor's accuracy. The field of view should, ideally, be 180° to cover the whole doorway area; however, smaller angles might be acceptable, depending on the size of the blind spots. The angular resolution is defined by the number of measurements along the field of view. From the acquisition perspective, α and the sampling frequency determine the vertical sampling resolution; β and the angular resolution determine the horizontal sampling resolution; and the raw sensor's accuracy determines the depth resolution. All of these parameters can be changed to provide different compromises between privacy and performance, especially if the number of users is small and excessive identification accuracy is not needed. The architecture of a physical sensor is cost-effective—the material costs for the prototype were under $100. It should also be noted that similar lasers are used in several home devices, including for small children to play with, as they are invisible to the human eye and pose no danger to humans or pets.

4.2. Self-Calibration and Input-Data Transformation

The transformation step is required to map the raw sensor data into the real-world Cartesian coordinate system. This enables the measured data to be used across different systems (e.g., across multiple sensors in the same household), regardless of the sensor type and its properties. The raw sensor data can be presented as a vector, where each value represents a distance at a specific angle (Figure 3, left). These data can be transformed into a coordinate system, as shown on the right of Figure 3.

The system can self-calibrate when no one is present. This can be done during the installation and repeated at any subsequent point, if needed. For self-calibration, we assume that the angle α is known and is fixed during the sensor's manufacturing. The sensor's height is then calculated using Equation (4), where α is the sensor's mounting angle and $L_{\beta/2}$ is the measured distance at the projected line midpoint, i.e., at $\beta/2$. The acquired depth with no one present is also recorded during the self-calibration stage to help with the users' body extraction at the acquisition stage.

During the acquisition process, all three Cartesian coordinates can be calculated using (3)–(5), where L is the measured distance, β' is the angle at which the distance was measured and β is the field of view.

$$Y_0 = L_{\beta/2} \cdot sin(\alpha) \qquad (2)$$
$$x = L \cdot sin(\alpha) \cdot cos(\beta' - \beta/2) \qquad (3)$$
$$y = Y_0 - L \cdot sin(\alpha) \cdot cos(\beta' - \beta/2) \qquad (4)$$
$$z = L \cdot cos(\alpha) \cdot cos(\beta' - \beta/2) \qquad (5)$$

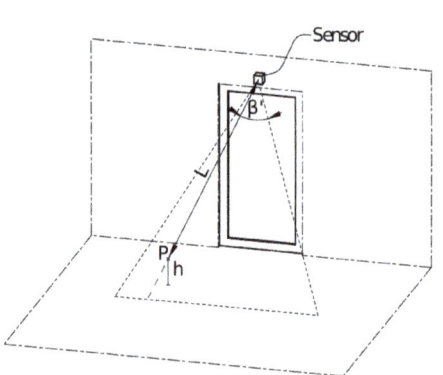 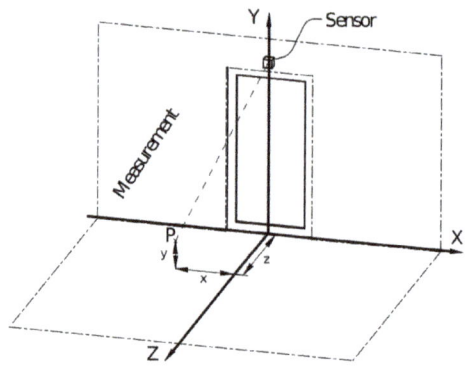

Figure 3. Transformation of the input data: (**left**) the depth measured with a one-dimensional depth sensor; and (**right**) the data transformed into the Cartesian coordinate system.

This kind of data transformation is valid, regardless of the type, technology and characteristics of a one-dimensional depth sensor. This allows the measured data or identification models to be shared between multiple sensors (e.g., in the same household). Sharing data through multiple identification sensors can be used to speed up the learning and consequently improve the accuracy. Therefore, with a data-transformation procedure, the identification accuracy of the whole sensor network can be improved.

4.3. Extraction of Features

To robustly identify the users, their body features need to be extracted from the transformed data. The measurements are pre-processed to remove the static background (e.g., ground and walls), leaving only measurements of the users for further processing. Next, for each non-background curve, five features are calculated. These features are used to determine users' crossing direction and identity.

The static background is removed from the data across the whole field of view by comparing the distance L of each measurement with the corresponding ground measurement, obtained during the self-calibration stage. Unless the measurement is significantly closer to the sensor than the background measurement, that data point is discarded. In this way, we obtain a *non-background curve*.

Next, five features are extracted from each non-background curve (Figure 4). The first feature (6) is the horizontal (x) distance between the first and the last curve's point, representing the measured object's width. The second feature (7) is the horizontal surface area (xz) between the curve and the shortest distance between the first and and last point, roughly corresponding to the measured object's volume. The third feature (8) is the maximum perpendicular distance from the shortest line from the first to the last point and the curve, representing the maximum object curvature. The last two features (9) and (10) are the maximum measured height and its horizontal position, which provides the user's location. In this way, each non-background curve is represented with these five features, which describe the user's body shape at a measured position.

$$F_1 = |x_0 - x_n| \tag{6}$$

$$F_2 = \sum_{i=1}^{n}(x_i - x_{i-1})|*|\frac{z_i - z_{i-1}}{2} - \frac{(z_n - z_0)(x_i - x_{i-1})}{2(x_n - x_0)} \tag{7}$$

$$F_3 = max\left(\frac{|(z_n - z_0)x_i - (x_n - x_0)z_i + x_n y_0 - z_n x_0|}{\sqrt{(z_n - z_0)^2 + (x_n - x_0)^2}}\right); \tag{8}$$

for $0 < i < n$

$$F_4 = max(y_i); \text{ for } 0 \leq i \leq n \tag{9}$$

$$F_5 = x(F_4) \tag{10}$$

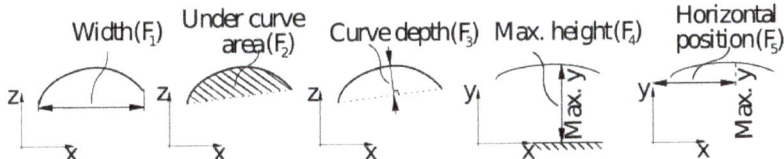

Figure 4. Graphical representation of the extracted features.

4.4. Acquisition Triggering and the Crossing Direction

The acquisition procedure is triggered when a non-background measurement (i.e., distances for the whole field of view) appears. The procedure stops when non-background measurements stop appearing. The acquisition yields a set of non-background measurements, which can represent one or more people crossing. To determine the number of people crossing between two background measurements, one or more rapid changes in the maximum height and/or horizontal position of the maximum-height features are observed, as shown in Figure 5D. A sudden drop and rise of maximum height feature indicates a new person is being measured. According to this logic, the input data are split into single person crossing events, which are later used to determine crossing direction and people identification.

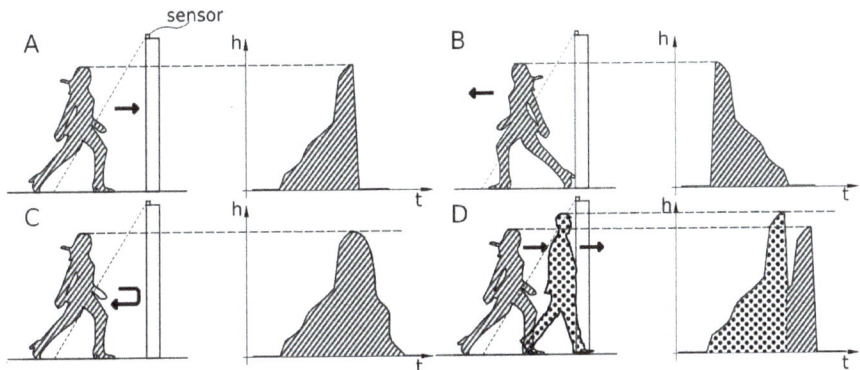

Figure 5. Response of the maximum-height feature (9) according to the crossing direction: (**A**) user walks toward the sensor; (**B**) user walks away from the sensor; (**C**) user first walks toward the sensor and then turns and walks away from the sensor; and (**D**) two users walk toward the sensor.

The crossing direction can be determined by observing the maximum-height feature. For each user, we keep a track of the maximum-height feature through the whole passing event. Due to the slope of the laser plane, the crossing direction can be determined, as shown in Figure 5. If the height rapidly increases and then slowly decreases (the laser beam encounters person's head first and the heels last), the user walks away from the sensor, as shown in Figure 5B; otherwise, the user walks towards the sensor. In this manner, both walking directions as well as other combinations (i.e., the user starts entering and then exits and the other way around) can be determined.

4.5. Identification

People identification is done at every crossing event. If multiple people are detected, it is performed once for each person. The identification is based on the processed body features acquired from a doorway crossing event, as shown in Figure 4.

In our framework, we do not assume a constant velocity of people; moreover, we allow for variations in the direction (e.g., as a consequence of hesitation). Therefore, the maximum-height feature (9) is used to re-order other acquired features (i.e., width (6), area under curve (7) and the curve depth (8)) along the temporal axis in the descending order of the maximum height. An example of these features is shown in Figure 6. The horizontal position feature (10) is not used in the identification, since it does not carry any relevant identity information. With sorted features, even if a user stops, moves slightly backwards and then continues the crossing, the features are still valid.

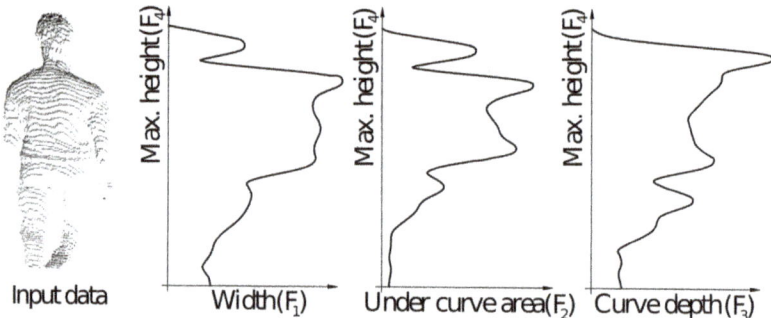

Figure 6. An example of input data and features used: width (6), curve depth (7) and area under the curve (8) sorted by the maximum-height feature (9).

Differences in the crossing velocity, which yield different amounts of data, are addressed by re-sampling the ordered features to the fixed number of samples. After re-sampling, a Fast Fourier Transformation (FFT) is used on each of the three features to obtain the Fourier descriptors (FD) [44]. From each re-sampled feature, 10 FDs are extracted (11). The maximum measured height is added to the 30 FDs, which gives 31 descriptors in total.

$$FD_{f,i} = |*| \frac{FFT(F_f, i)}{FFT(F_f, 0)}; \text{ for } 1 \leq f \leq 3 \text{ and } 1 \leq i \leq 10 \qquad (11)$$

Finally, a ML algorithm is applied to the descriptors to determine the user's identity. Since only one side of the body can be measured in a single crossing event (using one sensor), front and back ML models have to be built. The appropriate module is selected based on the user's walking direction. If the user walks away from the sensor, we assume that the back of the body is measured; otherwise, a model for the front of the body is used. The necessary assumption here is that the user is walking upright and forward.

5. Dataset

To evaluate the identification accuracy of the proposed approach, we used the described sensor setup to acquire a dataset containing the data from 18 subjects. For this task, a new, one-dimensional depth sensor prototype was developed specifically to meet all the requirements of our approach. The prototype sensor consists of Raspberry Pi B+ (Raspberry Pi Foundation, Cambridge, UK), NoIR camera module V1.3 (Raspberry Pi

Foundation, Cambridge, UK), 110° 780 nm laser line module H9378050L (Egismos, Burnaby, BC, Canada) and Acrylic NIR Longpass Filter AC760 (Midopt, Palatine, IL, USA). Laser module includes optics which transforms the beam into the 2D plane, thus enabling depth measurements along the line and ensuring safe intensity of laser light. However, any sensor fulfilling the application requirements can be used, regardless of the technology employed, e.g., a laser scanning device.

The developed one-dimensional depth sensor is based on a triangulation method with a depth-measuring range of between 0 and 3 m. The depth resolution depends on the measured depth and varies from 0 to 50 mm, as shown in Figure 7. An average depth of crossing in a real-life application depends on the height of the installation of the sensor and the height of an entering person—an adult comes closer to the sensor compared to a small child. However, in an average application and for an average-height person, the depth of the most important body parts such as head, shoulders and hips (the last one often covered by hands) should enable around 10 mm resolution, while for head it should only be around 5 mm on average. The sensor-sampling frequency is 30 Hz where 1240 measurements are taken simultaneously across the whole 120° field of view. These characteristics allow us to obtain a sufficient amount of data in a single doorway-crossing event (Figure 6). The sensor was mounted at an angle $\alpha = 60°$, and all the results reported were acquired with that geometry. Despite the relatively high resolution, the privacy-accuracy compromise could be adjusted simply by changing α.

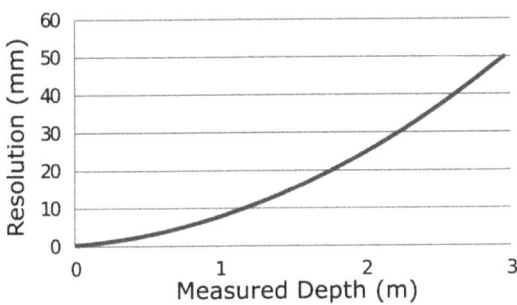

Figure 7. Sensor resolution with respect to the measured depth.

For the crossing direction detection and identification-accuracy evaluation, 670 crossing events involving 18 subjects aged between 21 and 47 were recorded. The identity and crossing direction of the subject in each passing event were recorded manually by an operator. The height of the smallest subject was 1.60 m and the height of the tallest was 1.94 m, whereas the average height of all 18 subjects was 1.76 m. The sample included 7 females and 11 males. The participants, all of whom were our colleagues at the institute, voluntarily signed written consent forms to conduct this research. At no time during the research was the health of the participants at risk since they only walked through a door and only a household common-type laser was applied on them. Both the privacy of the subjects and the confidentiality of the recorded data were maintained.

To obtain real-life scenarios, users' shoes and clothes (from a T-shirt to a jacket and a coat) were changed between the measurements. In addition, people carried everyday objects, such as umbrellas or backpacks, to make the conditions more realistic. In this way, a wide range of everyday scenarios was covered.

6. Experiments and Evaluation

To estimate the accuracy, a closed-set identification framework was used, i.e., the user was classified into one of the previously known identities representing, for example, the occupants of a household. This scenario is realistic for home use, where the number of occupants in a typical household is more-or-less fixed and adding another member is not a frequent event.

Overall, several experiments were performed to establish various properties of the proposed system in numerous scenarios. A couple of measurements are presented here. To determine how much data are needed to reliably identify people, two different approaches were explored. In the first approach, only the user's height was used, which is very privacy-preserving and easy to obtain. This approach is well known [45–47], and therefore served as our baseline. In the second approach, one-dimensional scanner data were used within our proposed framework, which might sound a bit privacy-invasive, but in reality it is not, and the additional data are essential to increase the identification accuracy. Both approaches were evaluated on the same dataset, but different amounts of acquired data were used to show the difference in the identification accuracy—only the maximum height in the first case and the full set of proposed descriptors when evaluating the proposed approach in the second case. In addition, another measurement was performed to evaluate which ML method performs best, but on a slightly different scenario.

6.1. Crossing Direction Detection

As described in the Acquisition Triggering and the Crossing Direction section, all the input data were split into single passing events by observing the maximum-height feature through the whole data acquisition process and then classified as if the person is walking in or out of the room. All 670 crossing events were classified correctly. The recorded crossings events were common for a home environment, i.e., only one person entering or leaving at the same time.

6.2. Identification by Subject's Height

It has been shown [45–47] that height can be used for the identification of smaller groups of people, whereas it is not suitable for larger groups, since sooner or later two users will be of equal height at least within the sensor's accuracy and under real-life circumstances. It should be noted that, in a real-life scenario, the measured user's height depends not only on the user's body measures but also on the shoes, haircut, head position and the speed or manner of walking. This makes an identification based on a single height measurement even harder and less robust in real life.

In the first experiment, the user's height is defined as the maximum height measured across a single doorway-crossing event (which is in our case $max(F_4)$). The distribution of measured heights is shown in Figure 8. Note that these measures could be obtained using single-point depth sensors, but we reused our depth data to calculate the subject's height for the purpose of this experiment.

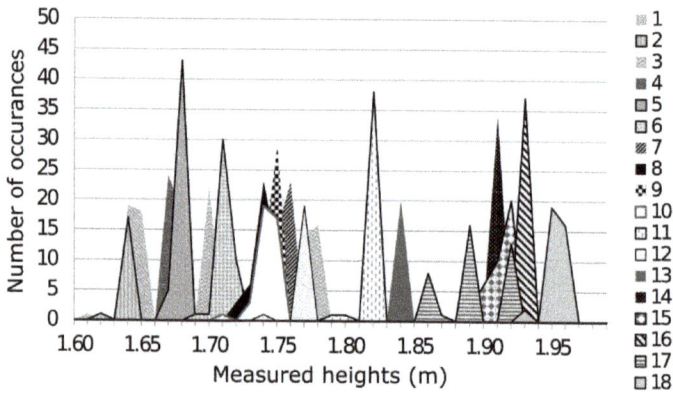

Figure 8. Distribution of heights measurements for the baseline experiment—identification by the subject's height. Different colors/textures denote different subjects.

The accuracy of the closed-set identification system based on height measures was estimated by using a nearest neighbor ML algorithm [43] and 10-fold cross validation [48]. The nearest neighbor algorithm classifies the person to the most similar person in the training set. To estimate how the size of a group affects the identification accuracy, the dataset was split into all possible combinations of groups of people. Next, each group of people was separately classified. The minimum, maximum and average closed-set identification accuracy with respect to the size of a group is presented in Figure 9 (lower curve). It is clear that the identification accuracy decreases with an increasing group size. Similarly, the maximum accuracy decreases, since larger groups of people are not diverse enough to properly identify all of the people with such a simple descriptor.

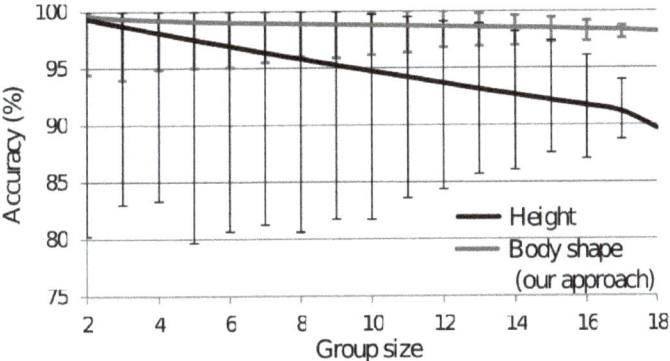

Figure 9. The minimum, maximum and average closed-set identification accuracy with respect to the size of a group. Whiskers denote the minimum and maximum value, as obtained by a 10-fold cross validation. The evaluation was made for different group sizes, from 2 to 18, taking into the account all the possible ways to construct groups of a certain size.

The results show that the closed-set identification accuracy using only the subject's height for all 18 subjects is 90%, which is quite better than described in [25], where the same accuracy was obtained for only three test subjects, but still not good enough for commercial use. We speculate that there might be several reasons for the relatively good results, e.g., not only better accuracy or position of the sensor, better sampling frequency or better height calculation, but also more consistent head cover, hair style or different shoe soles, as well as quite different heights of our randomly chosen test subjects. A much higher identification accuracy might not be achievable only by improving the sensor characteristics, since most of the noise comes from the environment. That is the rationale behind our approach—that AI methods are needed on redundant body measures to improve accuracy.

6.3. Our Approach: Identification by Subject's Body Shape

In this case, the input data are represented by 31 descriptors, as described in the Sensor-based Data Extraction section. Similar to the previous experiment, the closed-set identification accuracy estimation was made by using ML algorithms and a 10-fold cross validation for 18 subjects. Because of the larger number of descriptors, we tested several ML algorithms on the full dataset to establish the most appropriate one. The comparison of ML algorithms is shown in Table 1. A couple of the algorithms achieved similar very good performance, therefore there was no urgent need for testing or designing more algorithms at this stage, when the emphasis was on the sensor and the use of it. The best results were achieved by using the AdaBoostM1 ML algorithm [49], but, in terms of transparency, by a C4.5 classifier [50,51]—J48. Since a C4.5 classifier achieved such a good accuracy, it suggests that the 18 test subject were perfectly identified by the 31 descriptors if all data were captured and that additional noise such as object carried left enough descriptors intact to enable near-perfect identification. Using a full set of 18 subjects, AdaBoostM1 algorithm achieved 98.4% accuracy and is therefore our algorithm of choice. J48 (C4.5) is also an interesting choice particularly due to its simplicity, transparency and small number of required learning data, which is relevant for real-life applications. In the case of complex real-life domains, it might be more reasonable to opt for AdaBoostM1 since it usually achieves significantly better results than C4.5 [52]. Since Naive Bayes achieved only 87.76% accuracy, it indicates that the viewpoint through each single descriptor and then combined is less informative than the viewpoint through all of the descriptors combined at once.

Table 1. Comparison of ML algorithms in a 10-fold cross validation for 18 subjects.

Classifier	Precision	Recall	F-Measure	Accuracy [%]
AdaBoostM1 (J48)	0.98	0.98	0.98	98.36
LogitBoost (DecisionStump)	0.98	0.98	0.98	98.36
Bagging (J48)	0.97	0.97	0.97	97.31
LMT	0.97	0.97	0.97	97.16
IB1	0.97	0.97	0.97	96.87
J48	0.97	0.97	0.97	96.72
SimpleLogistic	0.97	0.97	0.97	96.57
RandomForest	0.96	0.96	0.96	96.12
BayesNet	0.95	0.94	0.94	94.33
MultilayerPerceptron	0.94	0.93	0.93	93.28
RandomTree	0.93	0.93	0.93	92.69
LibSVM	0.94	0.91	0.92	90.60
NaiveBayes	0.88	0.88	0.88	87.76

The results show a significant improvement in the average identification accuracy and its distribution compared to the height-only measurements, as shown by the minimum and maximum accuracy in Figure 9 (upper curve). The additional body-shape data in most of the cases enabled proper distinction of people of the similar height, which was the main purpose of using our approach. Furthermore, the narrower area of the identification-accuracy distribution suggests that this sensor is more reliable and robust than a sensor based on the user's height only.

6.4. Comparison of Theoretical and Practical Measurements

In Section 3, the theoretically computed accuracy under assumed noise and number of inhabitants is presented in Figure 1 to provide an initial estimate of which accuracy will be achieved under what circumstances when using a practical sensor setup. Moreover, the theoretical estimation allows an analysis on almost 4000 individuals and not only on 18 as in the physical measurements.

On the other hand, the practical measurements could show significantly different characteristics than the theoretical measurements, so a comparison between the two approaches is needed. In Figure 10, the experimental results marked with "our approach" are related to those from the simulation (without "our approach"). There are some similarities and some differences. In both approaches, whether theoretical or practical measurements, more body measurements improved the accuracy. When the same body measurements were compared, the theoretical results were slightly worse than the practical ones. The obvious reason for this is that there were 18 subjects with specific body sizes and nearly 4000 subjects in the database where, for example, many body sizes were similar and some were indistinguishably similar. As a result, the height comparison of the two approaches differs the most. Classification using only two physical measurements, e.g., height and shoulder width, already provided enough differences in the population to reduce the difference in accuracy between the physical and theoretical results. This is because the number of individuals with similar height is much smaller than the number of individuals with similar height and similar shoulder width. Since none of the theoretical measurements includes the full silhouette as in our approach, it was expected that the results of the practical measurements would be slightly better than the theoretical evaluations of the database, and indeed the comparison confirms this. However, the inclusion of four body measurements allowed an accuracy of about 97% for 10 inhabitants even in the theoretical measurements. In summary, the theoretical and practical measurements are sufficiently compatible to confirm the first hypothesis of our approach, once on 18 and the second time on almost 4000 individuals, from which combinations of 2–10 individuals were selected for the experiments.

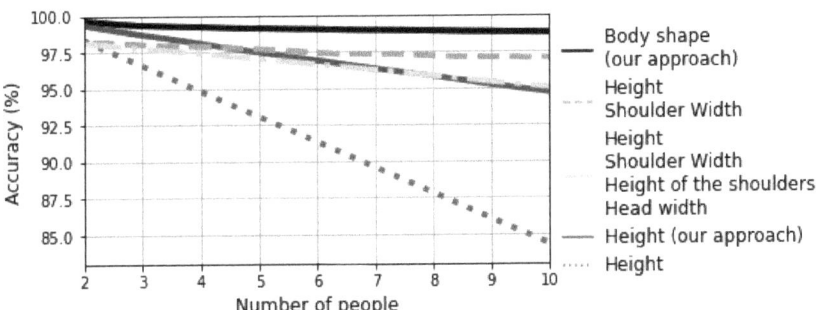

Figure 10. Experimental results along with theoretical identification accuracy according to different measures.

7. Discussion and Conclusions

In this paper, we present and test a sensor based on one-dimensional measurement for a novel approach to non-intrusive, non-invasive and privacy-preserving people identification in a smart environment, e.g., a household. Since the required accuracy in this AmI application is not as high as in security applications, and the number of people in a household is typically small, the proposed approach based only on body height or silhouette, e.g., body features, may prove sufficient for real-world applications.

Our results show that even the height-only approach achieves about 95% accuracy for nine tested individuals. However, the classification accuracy decreases rapidly as the number of people increases, dropping to 90% for 18 people, which means one error for 10 entries, which seems too high.

When the body shape approach is used instead of simple sensors that only measure body height, which requires the introduction of AI methods and more advanced computing, the accuracy is much higher and the results are more consistent (see Figure 9). The accuracy for 18 individuals remains above 98%, i.e., two errors in 100 entries. For five individuals, the accuracy reaches 99.1%. At the same time, the privacy intrusion remains practically nonexistent. Moreover, the ability to determine the crossing direction enables knowledge of the person's location down to the room level, as demonstrated in our experiments. With a cost-effective sensor implementation in terms of hardware and software, the laser-based approach, especially the body shape-oriented one, seems to be well suited for smart environments where data about the presence of people is needed but privacy must be maintained to a high degree. However, for high precision performance, practical application seems to be limited to families and small businesses with a small number of users.

The proposed identification system was evaluated in the laboratory on an apparent real-world scenario with 18 subjects, and the results were consistent with the preliminary study conducted using the ANSUR database. The live tests with our method on 18 subjects and the tests simulated with the ANSUR database on one million randomly selected groups of 3982 subjects consistently show that sufficiently good classifications can be obtained with the characteristic anthropomorphic data, e.g., height, shoulder width and hip width, for a small number of residents and under a reasonable noise. On the other hand, randomly selecting a small test group from the pool of nearly 4000 subjects more often results in individuals with similar characteristics, especially when the number of descriptors is small, e.g., height only. When multiple descriptors are considered, the differences between the practical experiments in the laboratory and the theoretical classifications from the ANSUR database fade away.

To further improve the detection performance, misidentified crossing events were manually checked. It was found that the sensor had problems with depth measurement in the presence of reflective clothing and different hairstyles. This problem could be addressed with an improved sensor design, which should then improve identification accuracy, or through improved AI methods such as Deep Neural Networks (DNNs). Nevertheless, both the worst and average identification accuracies remained high despite the depth measurement errors. Secondly, it was found that interference from another light source, such as sun reflections on the ground, made the results unreliable. However, this seems to be a problem for the type of sensor used in the experiments and can be avoided by using a different type of sensor or with additional filters.

For further work, advanced ML methods along with new feature extraction are being considered, including DNNs. However, such methods are not applicable in real life since the learning time of the system is short. Newer methods enable fast learning even with DNNs, but these new approaches need to be implemented and tested in detail. In addition, new methods for measuring one-dimensional depth are being tested to improve the overall identification accuracy. As shown in the preliminary study, the higher is the number and quality of features, the better is the identification accuracy. With further improvements, we believe it is possible to achieve face-recognition accuracy without violating privacy. In addition, extensive measurements are planned to test the performance with multiple people entering the room.

Finally, we present pros and cons of the proposed approach. The experiments strongly suggest that the laser-AI approach enables decent accuracy for real-world applications within the $100 target for a sensor; the system is reliable, non-intrusive, non-invasive and does not compromise user privacy. In addition, there is no comparable market solution to date.

However, in the cons category, there are a couple of obstacles. First, with five rooms in the department, it already requires $500, which can be somewhat of a dilemma. With massive use, the cost might go down to even $400, but it is yet to be achieved. In addition, each of the sensors requires power and either cables or batteries provide additional nuance. There are also competing solutions that are significantly cheaper than the laser system. An example would be a computer camera that fits into the $10 category. Such an approach introduces privacy concerns, as observed in several surveys, but there is still a certain percentage of users who trust their security systems to prevent capturing personal data for third parties. Finally, the benefit of knowing who is in a room may not be essential for ordinary residents, even though it is often declared as such in AmI publications. Most people either live alone or in fixed combinations in their own rooms such as bedrooms, and there the settings can be permanently matched with any motion sensor to alert the smart home that the room is no longer empty.

In summary, our approach based on a laser sensor and AI software enables sufficiently good accuracy to be used in real life, especially if users prefer non-invasive, non-intrusive and privacy-preserving systems for their smart home. The biggest concern is whether the cost–benefit ratio is indeed beneficial for current smart homes.

Author Contributions: Conceptualization, T.K. and J.P.; methodology, T.K. and J.P.; formal analysis, T.K. and J.P.; data curation, T.K. and J.P.; writing—original draft preparation, T.K. and J.P.; writing—review and editing, D.S. and M.G.; and supervision, M.G. All authors have read and agreed to the published version of the manuscript.

Funding: The authors acknowledge the financial support from the Slovenian Research Agency (research core funding No. P2-0209 and P2-0095).

Informed Consent Statement: Informed consent was obtained from all subjects involved in the study.

Data Availability Statement: The data collected during this study are available on request from the corresponding author. The data are not publicly available due to privacy reasons.

Acknowledgments: The authors would like to thank Eva Črnčič, Jernej Zupančič, Matej Trček, Tine Kolenik and Igor Gornik for their contributions.

Conflicts of Interest: The authors declare no conflict of interest.

References

1. Murshida, A.; Chaithra, B.K.; Nishmitha, B.; P B, P.; Raghavendra, S.; Mahesh Prasanna, K. Survey on Artificial Intelligence. *Int. J. Comput. Sci. Eng.* **2019**, *7*, 1778–1790. [CrossRef]
2. Cook, D.J.; Augusto, J.C.; Jakkula, V.R. Ambient intelligence: Technologies, applications, and opportunities. *Pervasive Mob. Comput.* **2009**, *5*, 277–298. [CrossRef]
3. Spoladore, D.; Mahroo, A.; Trombetta, A.; Sacco, M. ComfOnt: A Semantic Framework for Indoor Comfort and Energy Saving in Smart Homes. *Electronics* **2019**, *8*, 1449. [CrossRef]
4. Friedewald, M.; Elena Vildjiounaite, Y.P.; Wright, D. The Brave New World of Ambient Intelligence: An Analysis of Scenarios Regarding Privacy, Identity and Security Issues. In *Security in Pervasive Computing*; Clark, J., Paige, R., Polack, F., Brooke, P., Eds.; Lecture Notes in Computer Science; Springer: Berlin/Heidelberg, Germany, 2006; Volume 3934, pp. 119–133. [CrossRef]
5. Teixeira, G.D.; Savvides, A. A Survey of Human-Sensing: Methods for Detecting Presence, Count, Location, Track, and Identity. *ACM Comput. Surv.* **2010**, *1*, 1–41.
6. Akhlaghinia, M.J.; Lotfi, A.; Langensiepen, C.; Sherkat, N. Occupant Behaviour Prediction in Ambient Intelligence Computing Environment. *Int. J. Uncertain Syst.* **2008**, *2*, 85–100.
7. Spring, A. A History of Laser Scanning, Part 1: Space and Defense Applications. *Photogramm. Eng. Remote. Sens.* **2020**, *86*, 419–429. [CrossRef]
8. Spring, A. History of Laser Scanning, Part 2: The Later Phase of Industrial and Heritage Applications. *Photogramm. Eng. Remote. Sens.* **2020**, *86*, 479–501. [CrossRef]
9. Anton, M.A.; Ordieres-Mere, J.; Saralegui, U.; Sun, S. Non-Invasive Ambient Intelligence in Real Life: Dealing with Noisy Patterns to Help Older People. *Sensors* **2019**, *19*, 3113. [CrossRef]
10. Augusto, J.C.; McCullagh, P. Ambient Intelligence: Concepts and Applications. Computer Science and Information Systems. *ComSIS Consort.* **2007**, *4*, 1–27. [CrossRef]
11. Christen, M.; Bert Gordijn, M.L. *The Ethics of Cybersecurity*; Springer: Berlin/Heidelberg, Germany, 2020. [CrossRef]
12. Orr, R.J.; Abowd, G.D. The Smart Floor: A Mechanism for Natural User Identification and Tracking. In *CHI '00 Extended Abstracts on Human Factors in Computing Systems*; ACM: New York, NY, USA, 2000; pp. 275–276. [CrossRef]
13. Middleton, L.; Buss, A.; Bazin, A.; Nixon, M. A floor sensor system for gait recognition. In Proceedings of the Fourth IEEE Workshop on Automatic Identification Advanced Technologies (AutoID'05), Buffalo, NY, USA, 17–18 October 2005; pp. 171–176. [CrossRef]
14. Kalgaonkar, K.; Raj, B. Acoustic Doppler sonar for gait recoginition. In Proceedings of the 2007 IEEE Conference on Advanced Video and Signal Based Surveillance, London, UK, 5–7 September 2007; pp. 27–32. [CrossRef]
15. Adjabi, I.; Ouahabi, A.; Benzaoui, A.; Taleb-Ahmed, A. Past, Present, and Future of Face Recognition: A Review. *Electronics* **2020**, *9*, 1188. [CrossRef]
16. Arriaga-Gomez, M.; de Mendizabal-Vazquez, I.; Ros-Gomez, R.; Sanchez-Avila, C. A comparative survey on supervised classifiers for face recognition. In Proceedings of the 2014 International Carnahan Conference on Security Technology (ICCST), Rome, Italy, 13–16 October 2014; pp. 1–6. [CrossRef]
17. Tao, D.; Li, X.; Wu, X.; Maybank, S. General Tensor Discriminant Analysis and Gabor Features for Gait Recognition. *IEEE Trans. Pattern Anal. Mach. Intell.* **2007**, *29*, 1700–1715. [CrossRef]
18. Perš, J.; Sulić, V.; Kristan, M.; Perše, M.; Polanec, K.; Kovačič, S. Histograms of optical flow for efficient representation of body motion. *Pattern Recognit. Lett.* **2010**, *31*, 1369–1376. [CrossRef]
19. Dai, J.; Wu, J.; Saghafi, B.; Konrad, J.; Ishwar, P. Towards privacy-preserving activity recognition using extremely low temporal and spatial resolution cameras. In Proceedings of the 2015 IEEE Conference on Computer Vision and Pattern Recognition Workshops (CVPRW), Boston, MA, USA, 7–12 June 2015; pp. 68–76. [CrossRef]
20. Huynh, O.; Stanciulescu, B. Person Re-identification Using the Silhouette Shape Described by a Point Distribution Model. In Proceedings of the 2015 IEEE Winter Conference on Applications of Computer Vision, Waikoloa, HI, USA, 5–9 January 2015; pp. 929–934. [CrossRef]
21. Hasan, M.; Babaguchi, N. Long-term people reidentification using anthropometric signature. In Proceedings of the 2016 IEEE 8th International Conference on Biometrics Theory, Applications and Systems (BTAS), Niagara Falls, NY, USA, 6–9 September 2016; pp. 1–6. [CrossRef]

22. Gómez-Silva, M.J.; de la Escalera, A.; Armingol, J.M. Deep Learning of Appearance Affinity for Multi-Object Tracking and Re-Identification: A Comparative View. *Electronics* **2020**, *9*, 1757. [CrossRef]
23. Wu, L.; Wang, Y.; Shao, L.; Wang, M. 3-D PersonVLAD: Learning Deep Global Representations for Video-Based Person Reidentification. *IEEE Trans. Neural Netw. Learn. Syst.* **2019**, *30*, 1–13. [CrossRef] [PubMed]
24. Wang, S.; Zhang, C.; Duan, L.; Wang, L.; Wu, S.; Chen, L. Person re-identification based on deep spatio-temporal features and transfer learning. In Proceedings of the 2016 International Joint Conference on Neural Networks (IJCNN), Vancouver, BC, Canada, 24–29 July 2016; pp. 1660–1665.
25. Hnat, T.W.; Griffiths, E.; Dawson, R.; Whitehouse, K. Doorjamb: Unobtrusive Room-level Tracking of People in Homes Using Doorway Sensors. In Proceedings of the 10th ACM Conference on Embedded Network Sensor Systems, Toronto Ontario Canada, 6–9 November 2012; ACM: New York, NY, USA, 2012; pp. 309–322. [CrossRef]
26. Kouno, D.; Shimada, K.; Endo, T. Person Identification Using Top-View Image with Depth Information. In Proceedings of the 2012 13th ACIS International Conference on Software Engineering, Artificial Intelligence, Networking and Parallel/Distributed Computing, Kyoto, Japan, 8–10 August 2012; pp. 140–145. [CrossRef]
27. Kalyanaraman, A.; Hong, D.; Soltanaghaei, E.; Whitehouse, K. Forma Track: Tracking People Based on Body Shape. *Proc. ACM Interact. Mob. Wearable Ubiquitous Technol.* **2017**, *1*, 61, [CrossRef]
28. Munir, S.; Arora, R.S.; Hesling, C.; Li, J.; Francis, J.; Shelton, C.; Martin, C.; Rowe, A.; Berges, M. Real-Time Fine Grained Occupancy Estimation Using Depth Sensors on ARM Embedded Platforms. In Proceedings of the 2017 IEEE Real-Time and Embedded Technology and Applications Symposium (RTAS), Pittsburgh, PA, USA, 18–21 April 2017; pp. 295–306. [CrossRef]
29. Bassoli, M.; Bianchi, V.; Munari, I.D. A Plug and Play IoT Wi-Fi Smart Home System for Human Monitoring. *Electronics* **2018**, *7*, 200. [CrossRef]
30. Zou, H.; Zhou, Y.; Yang, J.; Gu, W.; Xie, L.; Spanos, C.J. WiFi-Based Human Identification via Convex Tensor Shapelet Learning. In Proceedings of the Thirty-Second AAAI Conference on Artificial Intelligence AAAI18, New Orleans, LA, USA, 2–7 February 2018.
31. Griffiths, E.; Assana, S.; Whitehouse, K. Privacy-preserving Image Processing with Binocular Thermal Cameras. *Proc. ACM Interact. Mob. Wearable Ubiquitous Technol.* **2018**, *1*, 133, [CrossRef]
32. Zhao, T.; Wang, Y.; Liu, J.; Chen, Y.; Cheng, J.; Yu, J. TrueHeart: Continuous Authentication on Wrist-worn Wearables Using PPG-based Biometrics. In Proceedings of the IEEE INFOCOM 2020—IEEE Conference on Computer Communications, Toronto, ON, Canada, 6–9 July 2020; pp. 30–39. [CrossRef]
33. Jin, H.; Su, L.; Ding, B.; Nahrstedt, K.; Borisov, N. Enabling Privacy-Preserving Incentives for Mobile Crowd Sensing Systems. In Proceedings of the 2016 IEEE 36th International Conference on Distributed Computing Systems (ICDCS), Nara, Japan, 27–30 June 2016; pp. 344–353. [CrossRef]
34. Jin, H.; Su, L.; Xiao, H.; Nahrstedt, K. Incentive Mechanism for Privacy-Aware Data Aggregation in Mobile Crowd Sensing Systems. *IEEE/ACM Trans. Netw.* **2018**, *26*, 2019–2032. [CrossRef]
35. Zhou, J.; Cao, Z.; Qin, Z.; Dong, X.; Ren, K. LPPA: Lightweight Privacy-Preserving Authentication From Efficient Multi-Key Secure Outsourced Computation for Location-Based Services in VANETs. *IEEE Trans. Inf. Forensics Secur.* **2020**, *15*, 420–434. [CrossRef]
36. Council Directive (EC). 1995/46/EC of the European Parliament and of the Council of 24 October 1995 on the protection of individuals with regard to the processing of personal data and on the free movement of such data. *Off. J. Eur. Communities L* **1995**, *1*, 40.
37. Langheinrich, M. Privacy by Design—Principles of Privacy-Aware Ubiquitous Systems. In *Ubicomp 2001: Ubiquitous Computing*; Lecture Notes in Computer Science; Abowd, G., Brumitt, B., Shafer, S., Eds.; Springer: Berlin/Heidelberg, Germany, 2001; Volume 2201, pp. 273–291. [CrossRef]
38. Hustinx, P. Privacy by design: Delivering the promises. *Identity Inf. Soc.* **2010**, *3*, 253–255. [CrossRef]
39. Gesto, M.; Tombari, F.; Rodríguez-Gonzálvez, P.; Gonzalez-Aguilera, D. Analysis and Evaluation Between the First and the Second Generation of RGB-D Sensors. *IEEE Sensors J.* **2015**, *15*, 6507–6516. [CrossRef]
40. Amara, K.; Ramzan, N.; Achour, N.; Belhocine, M.; Larbes, C.; Zenati, N. A New Method for Facial and Corporal Expression Recognition. In Proceedings of the 2018 IEEE 16th International Conference on Dependable, Autonomic and Secure Computing, 16th International Conference on Pervasive Intelligence and Computing, 4th International Conference on Big Data Intelligence and Computing and Cyber Science and Technology Congress (DASC/PiCom/DataCom/CyberSciTech), Athens, Greece, 12–15 August 2018; pp. 446–450.
41. Turan, C.; Wang, Y.; Lai, S.; Neergaard, K.D.; Lam, K. Facial Expressions of Sentence Comprehension. In Proceedings of the 2018 IEEE 23rd International Conference on Digital Signal Processing (DSP), Shanghai, China, 19–21 November 2018; pp. 1–5. [CrossRef]
42. Gordon, C. 1988 Anthropometric Survey of U.S. Army Personnel: Methods and Summary Statistics 1988, Final Report 1 Oct. 88–24 Mar. 89; Anthropology Research Project. Available online: http://mreed.umtri.umich.edu/mreed/downloads/anthro/ansur/Gordon_1989.pdf (accessed on 26 February 2021).
43. Cover, T.; Hart, P. Nearest neighbor pattern classification. *IEEE Trans. Inf. Theory* **1967**, *13*, 21–27. [CrossRef]
44. Zahn, C.T.; Roskies, R.Z. Fourier Descriptors for Plane Closed Curves. *IEEE Trans. Comput.* **1972**, *C-21*, 269–281. [CrossRef]

45. Srinivasan, V.; Stankovic, J.; Whitehouse, K. Using Height Sensors for Biometric Identification in Multi-resident Homes. In Proceedings of the 8th International Conference on Pervasive Computing, Helsinki, Finland, 17–20 May 2010; Springer: Berlin/Heidelberg, Germany, 2010; pp. 337–354. [CrossRef]
46. Godil, A.; Grother, P.; Ressler, S. Human Identification from Body Shape. In Proceedings of the 4th International Conference on 3D Digital Imaging and Modeling (3DIM 2003), Banff, AB, Canada, 6–10 October 2003; pp. 386–393. [CrossRef]
47. Ober, D.; Neugebauer, S.; Sallee, P. Training and feature-reduction techniques for human identification using anthropometry. In Proceedings of the 2010 Fourth IEEE International Conference on Biometrics: Theory Applications and Systems (BTAS), Washington, DC, USA, 27–29 September 2010; pp. 1–8. [CrossRef]
48. Hastie, T.; Tibshirani, R.; Friedman, J. *The Elements of Statistical Learning: Data Mining, Inference and Prediction*, 2nd ed.; Springer: Berlin/Heidelberg, Germany, 2009.
49. Freund, Y.; Schapire, R.E. Experiments with a New Boosting Algorithm. In Proceedings of the International Conference on Machine Learning, Bari, Italy, 3–6 July 1996; pp. 148–156.
50. Quinlan, J.R. *C4.5: Programs for Machine Learning*; Morgan Kaufmann Publishers Inc.: San Francisco, CA, USA, 1993.
51. Stuart, S.; Norvig, P. *Artificial Intelligence—A Modern Approach*; Prentice Hall Inc.: New Jersey, NJ, USA, 2003; pp. 697–706.
52. Gjoreski, M.; Janko, V.; Slapničar, G.; Mlakar, M.; Reščič, N.; Bizjak, J.; Drobnič, V.; Marinko, M.; Mlakar, N.; Luštrek, M.; et al. Classical and deep learning methods for recognizing human activities and modes of transportation with smartphone sensors. *Inf. Fusion* **2020**, *62*, 47–62. [CrossRef]

Article

Device-Free Crowd Counting Using Multi-Link Wi-Fi CSI Descriptors in Doppler Spectrum

Ramon F. Brena [1,*], Edgar Escudero [1,2], Cesar Vargas-Rosales [1], Carlos E. Galvan-Tejada [3] and David Munoz [1]

1. Tecnologico de Monterrey, School of Engineering and Sciences, Av. Eugenio Garza Sada 2501 Sur, Monterrey 64849, Nuevo León, Mexico; A00777396@itesm.mx (E.E.); cvargas@tec.mx (C.V.-R.); dmunoz@itesm.mx (D.M.)
2. Aerobit Technologies, Av. Eugenio Garza Sada 3820, Monterrey 64780, Nuevo León, Mexico
3. Unidad Académica de Ingeniería Eléctrica y Comunicaciones, Universidad Autónoma de Zacatecas, Jardín Juárez 147, Zacatecas Centro 98000, Zacatecas, Mexico; ericgalvan@uaz.edu.mx
* Correspondence: ramon.brena@tec.mx

Abstract: Measuring the quantity of people in a given space has many applications, ranging from marketing to safety. A family of novel approaches to measuring crowd size relies on inexpensive Wi-Fi equipment, taking advantage of the fact that Wi-Fi signals get distorted by people's presence, so by identifying these distortion patterns, we can estimate the number of people in such a given space. In this work, we refine methods that leverage Channel State Information (CSI), which is used to train a classifier that estimates the number of people placed between a Wi-Fi transmitter and a receiver, and we show that the available multi-link information allows us to achieve substantially better results than state-of-the-art single link or averaging approaches, that is, those that take the average of the information of all channels instead of taking them individually. We show experimentally how the addition of each of the multiple links information helps to improve the accuracy of the prediction from 44% with one link to 99% with 6 links.

Keywords: Wi-Fi; CSI; crowd counting; Doppler spectrum

1. Introduction

In recent years, mainly due to the COVID-19 health crisis in 2020 and beyond, the importance of technology capable of providing assistance to assess safety in crowds [1–4] has been brought to mainstream awareness [5]. However, crowd assessment applications are not limited to those that provide support for safety, and a new set of applications have been envisioned in businesses [6,7], and in other practical scenarios [8,9]. Of particular interest to the scientific community is the passive and device-free (meaning that the people who are monitored do not need to carry a device such as a cellular phone) estimation of the number of people in a given area. It is important to know the number of people in a room, to monitor human queues or to track the volume of customers in a commercial location, to provide valuable information in the context of smart space design, consumer marketing and venue security [10–12].

Though some recent and some decades-old developments have used computer vision for crowd measurement [1,13,14], nowadays visible light sensors are used with limitations due to the need of a line-of-sight which is subject to variable lighting conditions and coverage, as well as privacy concerns.

Recently, the increasing availability and descending costs of Wi-Fi equipment has promoted its use even in applications other than digital communications, such as indoor location [15,16]. In recent years it has begun to be the case of crowd measurement, given that popular Machine Learning techniques [17] can be used to recognize the disturbance patterns that human bodies produce when placed between a Wi-Fi transmitter and a

receiver [10]. Notwithstanding the wide range of potential applications that Wi-Fi sensing crowd analysis may reach, there are fundamental aspects of the subject (such as accuracy, reproducibility, and scale) that remain as limitations to overcome the boundaries of the current body of knowledge.

The aim of this work is to improve the results obtained from Machine Learning analysis of the disturbance patterns produced by human bodies to the signal propagation of individual channels of a Wi-Fi connection using the Doppler spectrum experienced in a crowd [18]. The original contribution of this paper is the systematical use of the Channel State Information (CSI) [19] of all the available channel links of a Wi-Fi communication (we refer to this as the 'multi-link' approach) rather than using indicators of just one link and discard the rest or to apply summary operations on all the links to reduce them to a single value (we refer to this as the 'single-link' approach, which is the one that has been used so far) to count the people present in one room, using a classification technique based on supervised Machine Learning. We demonstrate in this paper that the use of our multi-link approach improves accuracy in a dynamic environment with multiple wireless signals, multi-path components in the signal propagation through the channel that cause fading and absorption.

Many of recent works on this subject have used the Received Signal Strength Indicator (RSSI) as an index of the channel quality. RSSI is processed for feature extraction and the counting estimation is obtained after carrying out a learning phase [20–24]. For the estimation of the exact number of people in a place, the best RSSI-based reported results come from the work of Yoshida et al. [24], with a 77% of accuracy for up to 7 people. A major drawback of this technique is that RSSI-based algorithms tend to ignore the multi-path effects of the RF propagation, and as a matter of fact, its performance is greatly affected by channel disturbances. A more recent approach uses the Channel State Information (CSI) [19] that provides channel response information for Multiple Input Multiple Output (MIMO) Wi-Fi systems. As a result, CSI offers better measurements of people activity by capturing the disturbances the crowd cause in the channel.

Even with the use of CSI-based techniques, the performance of crowd-counting solutions documented in the literature present accuracy challenges that typically worsen with the group scale. The research carried out by Di Domenico et al. [25], which is commonly used as an indicator of the state of the art, reports an accuracy of about 80% for counting up to 7 people. The work from Xi et al. [26], is another common reference in the field, they achieved a probability of 80% of having either a perfect count or failing by one person when counting up to 30 people.

In this paper, we present a data driven work that takes advantage of the advances in Machine Learning (ML) techniques and apply it to multi-link Wi-Fi CSI information to produce better results than those reported in the literature for the recognition of the characteristics of a crowd, and specifically its counting.

The proposed method takes the information extracted from a CSI pattern of commercial off-the-shelf (COTS) Wi-Fi and translates it into the Doppler Spectrum where a set of features that capture information provided by the multi-link nature of the MIMO system is extracted to achieve high accuracy counting predictions. Furthermore, our approach can be potentially useful to identify dynamic characteristics of the crowd, such as its size, growth, dispersion and mobility, that could be applied to many relevant scenarios such as offering services based on the occupation detected in an environment, trends of influx in public spaces, occupancy predictions, mobility trends by region, and many more.

The method here presented reports several advantages with respect to other works such as:

- Fewer features derived from the signal are required for an accurate counting estimation. This results in reduced processing time since feature extraction requires less computing power.
- It works seamlessly with COTS Wi-Fi access points.

- Increased accuracy and other performance metrics as a result of using multiple links instead of one or an average.

The remaining of this paper is structured as follows. In Section 2 we present the background concepts for the sake of self-containment, as well as a review of the related work. Then, our method is presented in Section 3, together with the experimental setup and results description. Finally, in Section 4 we discuss the relevance of our contributions and provide some ideas for future work.

2. Background and Related Work

The field of crowd dynamics refers to the analysis of the motion of people within a defined group and its changes over a period of time. The topic has attracted an increasing interest from the research community due to many potential applications, and more recently the COVID-19 health crisis under way, makes clear that it is imperative to avoid crowd concentrations, especially in indoor spaces, in order to avoid further contagion [27]. Crowd applications are not limited to health issues though, and among other ones, we find crowd security and management for emergency handling, where the ability to recognize patterns in the crowd behavior allows better and faster responses or improvements to the space design [3,28,29]. Hence, several frameworks coming from different disciplines have been proposed in order to model the motion of a crowd.

The work of Helbing and Johansson [30], explores the analogies between the patterns of a crowd and the properties of a fluid of particles. Their study provides a framework to model the interactions of individuals in crowds, and the study of self-organized patterns of motion they generate as a result of the emerged collective intelligence and the social forces involved in the process. In a prior work, Helbing et al. [31], introduced the concepts of attraction and repulsion forces to simulate the dynamics of crowds in panic or evacuation situations.

In the following sections, it is also discussed how different authors put different levels of emphasis in one or more attributes of the crowd in order to model, describe and predict its dynamics; and each of them uses a set of metrics, either quantitative or categorical, as a basis for their work.

2.1. Quantitative Characterization of a Crowd

Still [32] defines crowd speed as 'the emergent speed of a group of individuals' that is a result of the non-linear interactions within the crowd in the local geometry. In his study, the author describes how crowd speed is modulated by crowd density (number of people by unit of area), being the flow volume a function of both.

Helbing et al. [33], analyze the relationships among numerical properties of crowds to describe a motion model. In this study, Helbing defines key crowd parameters such as density, speed, pressure and flow vectors. The research argues that even at highly dense crowds the motion of the crowd continues, which in turn causes dangerous 'turbulence' spots where crowd pressure is beyond a critical threshold.

The work of Pathan et al. introduces a novel approach for crowd behavior analysis and anomaly detection in coherent and incoherent crowded scenes [34]. The authors explore the crowd problem from a data-driven perspective and propose a method to calculate social entropy. The introduced metric is used as a descriptor of crowd behavior. Support Vector Machines are used to train and classify the flow feature vectors as normal and abnormal.

2.2. Categorical Metrics of Crowd Dynamics

Vicsek et al. [35], provide a general classification of collective motion. They proposed that any group of individuals can be categorized in five possible motion states: (i) disordered state (individuals moving randomly); (ii) fully ordered state (individuals moving at pace in the same direction); (iii) rotational (individuals moving in well-defined patterns); (iv) critical (state very sensitive to perturbations); (v) velocity correlated (individuals behave as elongated particles). Saleem et al. proposed a simplified approach to these categories by grouping them into coherent and incoherent crowds [34].

Interesting to our own work are crowd analysis studies that have been proposed in the image processing and computer vision fields. In this context, Xu et al. proposed an algorithm to detect the gathering and dispersing stages of a crowd in video recordings. To achieve this, the authors combine techniques of crowd counting and group entropy to estimate the spatial distribution of the individuals. These parameters are then used as features for the classification process [36].

2.3. Sensing Crowds with Wi-Fi

Human sensing based on commodity Wi-Fi has gained attention in recent years mainly because of the pervasive availability of Wi-Fi signals, that can be re-purposed sometimes with little cost. Advances in device-free sensing, where neither special equipment nor cooperation is required from the sensed subject, have been documented [37–39].

Amplitude and phase of Wi-Fi signals are very sensitive to the surrounded environment, and it has been shown that it is possible to extract patterns from such variations to identify human-related activity [23,40–42]. This way of acquiring human or crowd data is known as passive sensing. It refers to the family of sensing techniques that requires no cooperation from the sensed subject (as opposite to active sensing that is based, for example, in wearables or mobile apps used by the sensing target).

There are in the literature two different approaches to passive sensing with Wi-Fi signals: the first one is *passive device-present*, here the sensing signals from client-to-Access Point (AP) communication are exploited; the second one is *Passive device-free*, where sensing signals from AP-to-receiver communication are used. Passive device-free sensing based on commodity Wi-Fi has gained special attention due to the advantages of achieving crowd sensing without requiring the collaboration of the sensed group. Also, in contrast to device-present sensing, a device-free approach protects subjects privacy inherently.

2.4. Human-Centric vs. Crowd-Centric Sensing

Research work may also be classified according to the type of inputs it tries to detect: while human-centric aims to detect events and activities at the person level, crowd-centric aims to modeling properties of a group of people. Although a plethora of applications have been subject of study, in practical terms, sensing research may be binary classified in one of these two approaches.

As opposite to the human-centric perspective, crowd-centric sensing is not interested on predicting or estimating properties from single individuals, but those that arise from a group of people as an entity. Models in this category are designed to define crowd variables such as the size, density, speed and direction.

While human-centric sensing has found applications especially in health care [40,43–45], crowd-centric sensing has an important range of potential applications in public safety, mobility planning and marketing [6,46–50].

2.5. Sensing Crowd Properties

As discussed above in this section, crowds present several properties of interest that can be useful to prevent unwanted situations or stimulate desired behaviors. In this context, research work on Wi-Fi-based sensing models that can deliver accurate results is gaining attraction.

Wi-Fi-based crowd counting is the task of estimating the number of people gathered in a specific area. An increasing amount of research work has been dedicated to crowd counting with Wi-Fi signals in recent years. Xi et al. published a method to estimate the size of a crowd of up to 30 people using a Grey Verhulst model [26]. Another important crowd indicator is its density (i.e., the number of people per unit of area). While there is few documented research on this subject, a recent work by Tang et al. uses a device-present approach to calculate crowd density by capturing device's request probes and RSSI signal [51]. Depatla et al. [52] proposed a framework to sense speed of a crowd in both outdoor and indoor locations.

2.6. Rssi and Csi: The Sensing Signals

Wi-Fi is a commercial denomination for the IEEE 802.11 standard, which is the predominant technology used for Wireless Local Area Networks (WLAN) that operates in the 2.4 GHz or 5 GHz frequency bands. The first version of the standard was released in 1997 [53]. A decade later, the idea of using Wi-Fi signals to identify, analyze or predict human activity started to be increasingly present in researchers' work desks around the world. Since then, several techniques have been proposed for capturing, denoising, processing and classifying the hints of human activity that are intrinsically carried by the wireless signals the Wi-Fi standard uses.

Estimation techniques are in general based on some kind of measurement of signal parameters on the RF propagation channel as a function of the variables of interest. There are two specific types of signal-related parameters that are commonly used in current Wi-Fi sensing research, which are: RSSI and Channel State Information (CSI).

2.6.1. Rssi

Due to the simplicity of field strength measuring, RSSI has been commonly used in many applications as those concerned to indoor localization. As its name describes, RSSI is a measure of the power present in the signal at the time it arrives to the receiver. According to signal propagation models and measurements, signal energy decreases with distance. Because of this, RSSI is often used with multilateration methods to estimate position [15,16]. This is a device-assisted approach, as it requires that the subject to be localized carry a device with a Wi-Fi receiver. Human tracking with Wi-Fi can, however, also be accomplished by device-free methods. This is achieved by analyzing the variance and other statistical properties of the RSSI signal data [54,55]. Youssef et al. for example, used the moving average of the signal strength values to track the location of a person [56].

Although RSSI can be easily implemented, its suitability as sensing vehicle in non-controlled environments is limited, as RSSI presents several impairments when obstacles are present in the area of interest. The strength of a single received signal is greatly affected by multipath and shadowing effects that yield estimation errors. As a result of this, CSI has been proposed as an alternative [57].

2.6.2. Csi

Modern Wi-Fi standards, powered with Orthogonal Frequency Division Multiplexing (OFDM) modulation and MIMO capabilities, utilize CSI as an indicator of the properties of the channel to dynamically optimize and adapt the transmission parameters to improve performance. Therefore, CSI is an overall representation of the channel state that sums up the signal's propagation effects including scattering, fading and multipath [58]. Although RSSI also offers an overall picture of the channel by providing time averaged total power of the signal envelope at the receiver, CSI is an estimation of the channel coefficients that represent either the impulse or the frequency response at the sub-carrier level [59]. Because of its granularity of sub-carrier frequencies and its vector representation, CSI data provide more information of the channel impairment effects that the signal experiences in contrast to the received power strength given by RSSI.

The proposed research work makes use of the advantages mentioned above and employ CSI as the sensing signal. Hence, it is worth providing a more detailed description of the nature of the CSI data and the context in which it is produced.

2.6.3. Mimo Systems

In wireless communications the information signals are transmitted through a channel, ideally in line-of-sight (LOS) obstacle-free conditions, but in practical scenarios, the LOS condition is not met and the transmission paths are not unique, so the beam uniqueness does not hold. The objects and subjects located in the surroundings of the channel may reflect, refract, diffract and scatter the signal, producing multiple new paths the signal traverse to the receiver, an effect called multipath propagation that exhibits deep fades with high variance [60].

From the receiver point of view, multiple copies of the signal arrive with different time delays, amplitudes and phase shifts. This aggregation of signal replicas produces fading, a multipath-induced interference that results in variations of Signal-to-Noise Ratio (SNR) of the emerging signal [61]. Such a behavior is traditionally considered as a problem the wireless communication system has to solve.

A Multiple Input Multiple Output (MIMO) system uses multiple antennas at the transmitter or the receiver to improve the overall communication performance. In contrast to single-antenna systems (SISO) that gather a single signal for reception, MIMO systems are able to use simultaneously multiple signals carrying information. Then, signal processing techniques such as space-time coding, beamforming, channel estimation and symbol detection are used to take advantage of the various signals that are available at the receiver. This results in significant improvements of spectral efficiency, data rate and system capacity [62].

A typical configuration of a MIMO channel consisting of M number of antennas at the transmitter and N antennas at the receiver is shown in Figure 1. The transmission is expressed in terms of the received vector **y**, the transmitted vector **x**, the channel coefficient matrix **H** and noise **n**, as follows:

$$\mathbf{y} = \mathbf{H}\mathbf{x} + \mathbf{n} \qquad (1)$$

Similarly, each of the N antennas at the receiver, receives signals from all the M transmitting antennas, creating S number of communications links according to:

$$S = MN \qquad (2)$$

The general expression for (1) in its matrix representation is

$$\begin{bmatrix} y_1 \\ y_2 \\ \dots \\ y_N \end{bmatrix} = \begin{bmatrix} h_{11} & h_{12} & \dots & h_{1M} \\ h_{21} & h_{22} & \dots & h_{2M} \\ \dots & \dots & \dots & \dots \\ h_{N1} & h_{N2} & \dots & h_{NM} \end{bmatrix} \begin{bmatrix} x_1 \\ x_2 \\ \dots \\ x_M \end{bmatrix} + \begin{bmatrix} n_1 \\ n_2 \\ \dots \\ n_N \end{bmatrix} \qquad (3)$$

For example, a 3 × 2 MIMO system will have 3 transmitting antennas and 2 receiving antennas, conforming a total of 6 communications links. Each link's channel coefficient $h_{m,n}$ represents the channel effects that the signal that travels to the m-th receiving antenna from the n-th transmitting antenna undergoes. Thus, for the 3 × 2 example, the received vector is represented as follows:

$$\begin{bmatrix} y_1 \\ y_2 \end{bmatrix} = \begin{bmatrix} h_{11} & h_{12} & h_{13} \\ h_{21} & h_{22} & h_{23} \end{bmatrix} \begin{bmatrix} x_1 \\ x_2 \\ x_3 \end{bmatrix} + \begin{bmatrix} n_1 \\ n_2 \\ n_3 \end{bmatrix} \qquad (4)$$

Notice that, as previously mentioned, a CSI packet is a complex number representing the amplitude and phase of the channel state; so in MIMO-enabled Wi-Fi, CSI signals provide the estimated values of the channel coefficient matrix **H**.

2.6.4. Ofdm Transmission

The IEEE 802.11 standard adopted Orthogonal Frequency Division Multiplexing, or OFDM, as part of its transmission technique to achieve higher data rates [62]. The idea behind this technology is to increase data rate using parallelism by dividing the assigned spectrum into several narrowband sub-carriers that are used for simultaneous transmission. The OFDM sub-carriers are orthogonal in the mathematical sense, which implies that sub-carrier frequencies are selected to cancel out inter-symbol interference (ISI). Therefore, additionally to delivering higher rates, OFDM offers immunity to the ISI effect caused by multipath fading and it also requires relatively simple receivers to reconstruct the transmitted data, since signal processing is done using the Fast Fourier Transform (FFT) and the inverse FFT algorithms instead of hardware.

According to standard IEEE 802.11n, MIMO data are modulated into 52 sub-carriers using Inverse Fast Fourier Transformation (IFFT) and transmitted as OFDM symbols in discrete packets. The receiver measures the CSI for each packet and adapts parameters to channel variations, the received signal is then demodulated by applying the direct FFT. The tool provided by Halperin et al. [59], which was used to obtain our experimental data, allows the extraction of CSI information from an Intel 5300 wireless card, exposing channel information of 30 of the 52 Wi-Fi sub-carriers. The CSI for each sub-carrier is defined as

$$h = |h| \, e^{j\theta} \qquad (5)$$

where $|h|$ and θ represent the magnitude and phase of the communication channel, respectively.

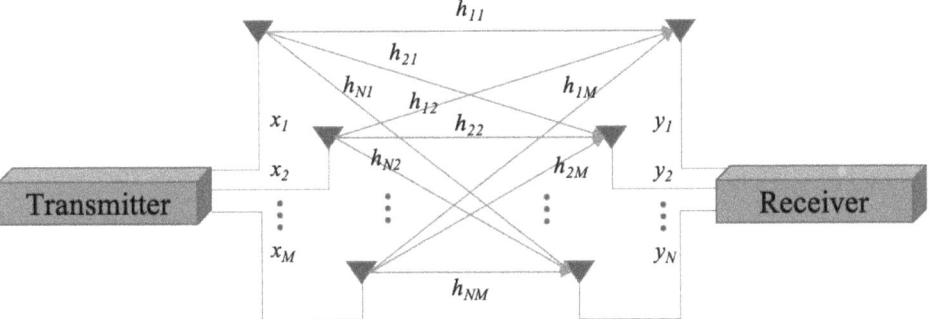

Figure 1. Multiple Input Multiple Output System.

2.7. Related Work

In this section we briefly review relevant literature that documents the current state-of-the-art of device-free Wi-Fi-based crowd counting. Each subsection corresponds to a published article in the field. In the last subsection we provide a summary table with the reported accuracy of each reviewed method for further reference and benchmark.

2.7.1. Trained-Once Device-Free Crowd Counting and Occupancy Estimation Using Wi-Fi: A Doppler Spectrum Based Approach

The work from Di Domenico et al. [25], performs Doppler Spectrum transformation to a CSI stream. The authors carried out a series of experiments where groups of people with several participants (from 0 to 8) where sensed in 3 different locations. Due to the scope of the experiments, the resultant dataset is of enormous value for the research community and it is the one we use for the preliminary work of the present research.

Di Domenico also introduces a long list of features that can be extracted from the Doppler spectrum matrices. Among all the possibilities the authors selected Spectral Kurtosis as the unique descriptor for their model. The performed series of arithmetical mean to the sub-carriers and links in order to get to a simpler parameter to work with.

Also, for the learning stage, the authors used a Naive Bayes classifier, because of its simplicity as a probabilistic algorithm. With this setup, they achieve about 80% of accuracy for crowd counting estimation. It is worth noticing that the main purpose of the work by Di Domenico et al. is to show the advantages of their method for a 'training once' scenario where there is no need for dedicated training in every different location.

2.7.2. Frog Eye: Counting Crowd Using Wi-Fi

The article from Xi et al. [26], is another often cited work in the Wi-Fi-based sensing field. In this research the authors documented a sensing framework based on CSI measurements from off-the-shelf Wi-Fi equipment.

Xi's model introduces a feature called Percentage of non-zero elements (PEM), which is a measurement technique based on the non-zero counting of the CSI dilated matrix. The resulting dataset is classified with the help of a grey Verlhust model factor.

To measure the performance of their method, the authors utilized the probability that an error equal or less than a defined threshold occurs for a particular counting estimation. This indicator was reported to be 98% for a threshold of 2 or fewer person and about 80% for a threshold error of 1 person.

2.7.3. Wicount: A Deep Learning Approach for Crowd Counting Using Wi-Fi Signals

The work from Liu et al. [63,64], explored the capabilities of a deep learning-based method for crowd counting. It uses a fully connected neuronal network with two hidden layers. It also implements regularization and exponential decay to improve performance. The experimental results show that the introduced deep learning model is able to estimate the number of crowd up to 5 with the accuracy of 82.3%.

The key contribution of the article for the Wi-Fi sensing field is that it documents a Deep Learning model that is arguably the first in its kind to be applied for crowd counting. Even if one can claim that the amount of time and computing resources required to train a DL system are still very demanding and the outcome quality does not correspond to the effort, the authors clearly pointed out a direction for future work. Similar works with data coming from Wi-Fi CSI information, which use Deep-Learning, like Cheng et al. [65], achieve slightly higher accuracy, with a reported 88.66%.

2.7.4. Occupancy Estimation Using Only Wi-Fi Power Measurements

The article from Depatla et al. [20] introduced, to the best of our knowledge, the most cited RSSI-based method for crowd counting. Their framework is based in a model that incorporates the pattern of both blocking LOS and scattering that human bodies produces in the strength of the Wi-Fi signals.

The authors approached the problem from an analytical perspective to obtain a mathematical expression that relates the signal strength with the PDF of the number of people in a crowd. Then, the method uses Kullback-Leiber divergence to estimate the size of a crowd with up to 9 people.

As in other works in the literature, Depatla presented its results as probabilities of getting errors of certain number of people. Specifically, their method reported $P(e \leq 1)$ = 55% and $P(e \leq 2)$ = 63% for indoor experiments with off-the-shelf equipment. For the outdoor scenario $P(e \leq 1)$ was 64%, while achieving 96% for a threshold of 2 or less people of counting error.

2.7.5. Estimating the Number of People Using Existing Wi-Fi Access Point in Indoor Environment

Yoshida et al. [24] published a relevant work where the counting estimation is made using regression algorithms. The method uses RSSI data as feature and test both linear regression and and SVM regression to make the classifications.

The experiment setup consisted in a single transmitter and four receivers, each of them working as independent measuring point. A notable contribution of this research is that it explores, with this kind of layout, additional crowd characteristics such as density and presence/absence of people.

The accuracy rate of Yoshida's method for estimating the number of people is 77%, for estimating the degree of congestion (crowd density) is 95%, and for estimating the presence/absence of people is 98%. This work was updated by Mabuchi [66] to achieve 93% in counting smalls groups of people.

2.7.6. Freecount: Device-Free Crowd Counting with Commodity Wi-Fi

FreeCount [67] is a high-accuracy system for crowd counting that uses a set of features of three kinds: statistics, frequency domain and shape. In their publication, Zou et al. focus in the problem of temporal variation of the channel conditions and the unpractical need of re-training the classifier every period of time.

By using a model based on SVM, the author reported a crowd counting accuracy of 99%, and $P(e \leq 1)$ = 97%. Moreover, FreeCount implements transfer kernel learning (TKL) to cope with the changes in the channel condition with time. With TKL as the SVM kernel FreeCount reported an accuracy of 96% two weeks after the actual trainning took place.

A downside of the FreeCount approach is that it requires to modify the Access Points in the location. This means the solution is not "commercial off-the shelf", in the sense that it can not work seamless with currently installed infrastructure as the rest of the methods reviewed above can. For this reason we considered it a non-COTS solution.

Table 1 shows a summary of the performance of crowd counting methods that use COTS through Wi-Fi technology

Table 1. Performance summary of crowd counting methods using commercial off-the-shelf (COTS) Wi-Fi.

			Accuracy			Scale	
Author	Data Source	Classifier	Accuracy Rate	$P(e<1)$	$P(e<2)$	Max # of People	Sensing Area (m²)
Di Domenico	CSI	Naive Bayes	81%	-	91%	8	30, 45, 70
Xi	CSI	Grey Verhulst	-	80%	98%	30	-
Liu	CSI	Deep Learning	82%	-	-	5	-
Depatla	RSSI	Math. Exp.	-	55%	63%	9	33
Yoshida	RSSI	SVR	77%	-	-	7	-

2.8. Theoretical Framework of Crowd Characterization with Wi-Fi Csi in the Doppler Spectrum

In a real propagation environment, a signal propagates along multiple paths, and the receiver experiences multiple time-delayed replicas of the transmitted signal. Furthermore, if the receiver is moving, a set of Doppler shifts occurs in the receiving end and a Doppler spread spectrum arises. In a MIMO-OFDM transmission, random variations on the sub-carriers frequency causes uncorrelated fading between the different received paths. If a simple correlation receiver is applied to the received signal, delayed versions of the transmitted signal will not correlate properly and thus cause self-interference [47].

For multipath communication, the radio signals arrive at the receiver device as the sum of all the contributions produced by the scattering process. When the scatter objects are static with respect to the radio source, the radio frequencies do not change in the propagation channel. However, in the case of scatterers or source motion, there will be a Doppler shift that depends on the speed and moving direction with respect to the signal propagation path. At a single frequency level this phenomena is known as Doppler shift, but in a time-varying scenario (as the scatter objects change direction and speed over time) a set of Doppler shifts or Doppler spread is also referred to as "Doppler spectrum".

The work of Yang et al. [68], provides a conceptual framework to analyze the Doppler spectrum of a Wi-Fi transmission using the CSI signals. We will briefly summarize Yang's analytical model in the following lines for convenience. The scenario is illustrated in Figure 2 and is described as follows: if a transmitter is at a distance d from a moving receiver with velocity v at a given instant, then the Wi-Fi signal that will arrive at the receiver is affected by a channel that is multipath and time-varying. Let's suppose there are a total of L independent paths l, each of them with different angles of arrival θ_l to the receiver's moving direction.

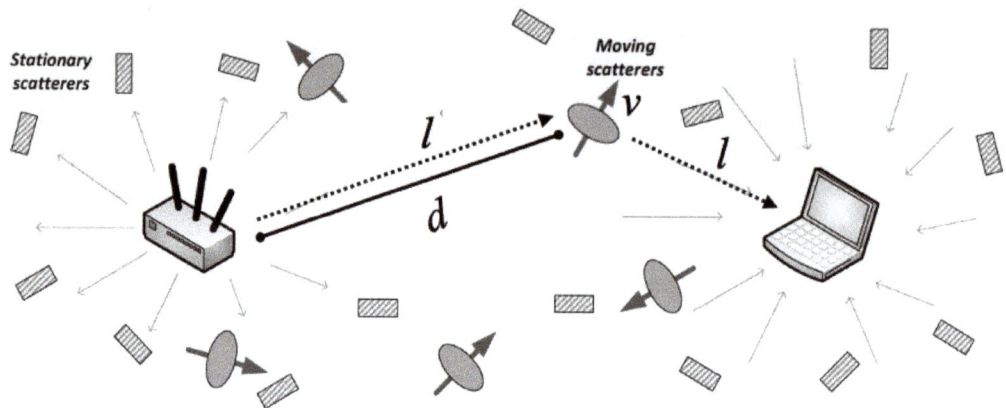

Figure 2. Doppler shifts produced by moving scatterers in a Wi-Fi link.

In MIMO-OFDM Wi-Fi there are k sub-carriers and multiple links that will be affected by the same multipath process, the channel response can be expressed as a function of the time instant nT_s

$$h_k(nT_s) = \sum_{l=1}^{L} \alpha_l e^{-jw_k \frac{d_l}{c}} e^{jw_k \frac{v_l}{c} nT_s} + \epsilon_k \tag{6}$$

where α_l is the amplitude of the l-th path, c is the propagation velocity of electromagnetic wave, ϵ_k is the measurement error, and $w_k \frac{v_l}{c}$ is the Doppler shift.

From the Wi-Fi IEEE802.11n standard [69], we know that the center frequency for each OFDM sub-carrier k in GHz is given by

$$f_k = 2.4 + k\Delta f \quad k = 1, 2, ..., 30 \tag{7}$$

where Δf is the frequency difference between sub-carriers, and it is extremely small compared with the 2.4 GHz. The Doppler shift can be approximated as

$$w_l = 2\pi \frac{v_l}{c} \tag{8}$$

Hence, a good approximation of the channel response can be given by the average of the available individual channel responses of the subcarriers, as follows:

$$\bar{h}(nT_s) = \frac{1}{30} \sum_{i=1}^{L} \sum_{k=1}^{30} \alpha_l e^{-jw_k \frac{d_l}{c}} e^{jw_l nT_s} + \bar{\epsilon_k} \tag{9}$$

Next, we obtain the frequency domain channel response by using the discrete Fourier transform as follows:

$$\bar{H}(m) = \sum_{n=1}^{N-1} \bar{h}(nT_s) e^{-j(\frac{2\pi}{N})nm}$$
$$= \frac{1}{30} \sum_{i=1}^{L} \sum_{k=1}^{30} \alpha_l e^{-jw_k \frac{d_l}{c}} \frac{1}{30} \sum_{k=1}^{30} e^{j2\pi(f_l T_s - \frac{m}{N})n} + \bar{E_k} \tag{10}$$

Equation (10) is an analytical representation for Doppler Spectrum of WiFi CSI. From this foundation, different authors take different approaches for crowd counting.

For instance, Zou et al. [67], use a set of features coming from statistics of magnitude and phase, Fourier transformation and shaped metrics, all combined to achieve predictors of human motion. Di Domenico et al. [25], extract Average Spectral Kurtosis as the unique feature they use for their model, while Yang et al. [68], use only the first link of the MIMO grid for feature extraction.

3. Proposed Method and Results

The method here presented is based on the hypothesis that the diversity in channel response information that multiple communication links of a MIMO system carry could provide better descriptors of the number of people in a crowd than a single channel or a channel average. While other authors average or discard the multi-link information [25,68], our method exploits such information to produce high-quality features for the estimation model.

The results presented in this paper follow a data-driven, quantitative approach. We show how the data can be processed to get estimations of the crowd characteristics with acceptable accuracy. The methodology followed in this paper includes:

- The use of a reference dataset (see Section 3). On using a public dataset instead of collecting our own experimental data, we give up the possibility of adding information we might find useful. However, there is the opportunity to directly compare our results to the ones obtained by other researchers.
- Implementation of our method in MATLAB, and the further training of the model with a set of learning algorithms.
- Performance assessment of our method in terms of accuracy and other results providing quality indices, and a comparison with state-of-the-art approaches by re-implementing them in order to have a reliable comparison.

A high-level view of our method is illustrated in Figure 3. The data collection (which we took from the public dataset of Di Domenico et al. referenced before) is on the dotted box to the left, and our process appears inside the right dotted box. Our data-driven method comprises the steps:

1. Doppler spectrum estimation: in this step we created a MATLAB script that transforms the available CSI data points to the spectrum domain in order to obtain Doppler spectrum data as described in Equation (10).
2. Feature extraction: From the CSI readings together with the Doppler spectrum estimated parameters, a vector of signal features is derived, which is supposed to be a good compact representation of the signal characteristics, at least for the classification purposes that we have, that is, the estimation of the number of people in the room. We used mainly statistical descriptors of the signal, which could be a good or bad idea, and we can only assess this later on, when we obtain the classification performance figures. The output of the figure extraction phase is a *dataset*, which is a table in which rows are individual observations and columns are the features. The dataset is the starting point of the Machine Learning process itself.
3. Train-test split: In order to assess the prediction quality of the trained Machine Learning classifier in a fair way, we need that the data used for testing its performance has not already seen by the classifier; so we make a separation or *split* of the dataset into two subsets: the *training* dataset and the *testing* one. The relative size of each one is critical for a good performance, and this will be discussed below in the corresponding subsection.
4. Classifier training: Using the training part of the dataset, we adjust the parameters of a standard classifier (like Random Forest and others, described below). All of those classifiers are readily available in programming libraries, for the different platforms (MATLAB in our case), so the real work is not to construct the classifiers but to choose and configure them properly; whether or not it has been well done is only seen later, when classification performance results are obtained.

5. Prediction assessment: The already trained classifiers are used to obtain, for each row in the testing part of the dataset, a predicted class (in this case a number of people in the room). Once the prediction is done, its quality can be assessed by a number of well-known metrics such as accuracy, precision, recall and others, which will be discussed below when we present experimental results.

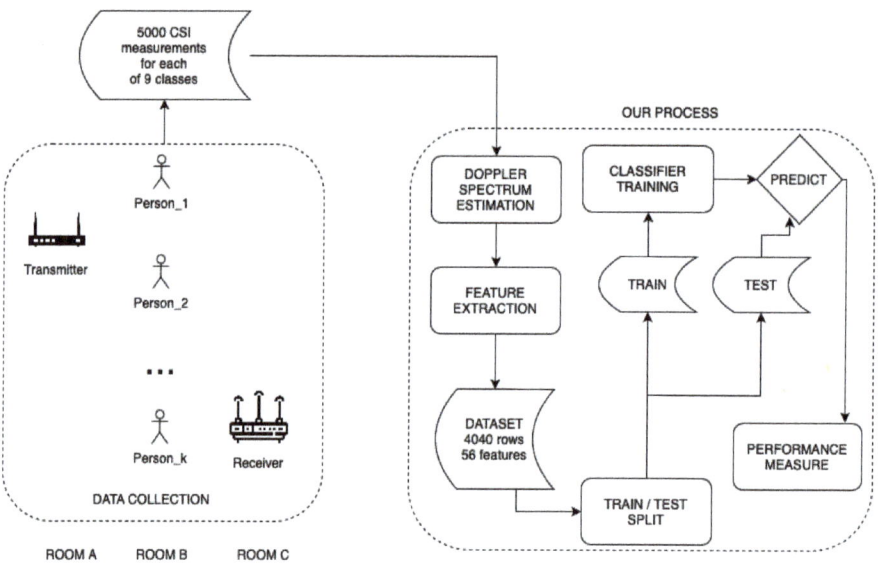

Figure 3. High-level diagram illustrating our method.

From the 4040-row features dataset to the classification assessment, everything is a mostly standard Machine Learning process (for which there are even free ML software platforms), so we do not claim to make any contribution there. Our contribution is the process that goes from the instrumentation readings, available from the Di Domenico et al. public data base [70] to the dataset construction from which the ML process is done, but of course the usefulness of the Doppler spectrum information, as well as the features we proposed can only be seen once the predictive power of the trained classifier is fully assessed.

3.1. Multi-Link Based Csi Crowd Counting Estimation

Our model calculates a set FS of p feature vectors F, each one of them being a combination function g_i of a descriptor $d_i \in D_i$ applied to each one of the T_{lk} available links $\overline{H}_{lk}$ in the MIMO Wi-Fi transmission. From the set of feature vectors our goal is to estimate, for a given one not previously seen, the number of people in the room with an accuracy as high as possible.

This set of feature vectors can be represented as follows:

$$FS = \{F_0, F_1, ..., F_p\}, \quad \text{where} \quad F_i = g_i(D_i) \quad for \quad i = 0, 1, 2, ..., p \\ and \quad D_i = \left\{ d_i(\overline{H}_{lk_1}), d_i(\overline{H}_{lk_2}), ..., d_i(\overline{H}_{lk_{T_{lk}}}) \right\} \tag{11}$$

The exact way of defining the features for Machine Learning prediction is entirely domain-dependent, and some argue that it is as much an art as is a science. In this paper, we explore the prediction performance of the model when d_i are standard statistical measures.

3.2. Dataset

In order to experimentally test the performance of our multi-link method, we used a publicly available dataset. The dataset from Di Domenico et al. [19] provides great opportunities to explore and test our Wi-Fi-based crowd counting hypothesis in a data-driven way. This dataset consists of the following:

- A 2-antenna Wi-Fi transmitter (an off-the shelf AP) and a 3-antenna Wi-Fi receiver (a computer with an Intel 5300 NIC) are set in the experiment location.
- Groups from 1 to up to 8 people were sensed using the mentioned setup, additionally to the 'empty room' case.
- The volunteers are allowed to move freely, but the only meta-data being labeled is crowd counting.
- A CSI trace is extracted every 20 ms, with a round lasting about 2 min for every counting case.
- The whole experiment was repeated in 3 different locations, as follows: Room A is a small size office room (5 m × 6 m), Room B is a medium size meeting room (5 m × 9 m), and Room C is a large size meeting room (6 m × 12.5 m).

The Di Domenico's dataset includes at least 5000 CSI measurements for each counting class, for each type of Room. Also, each CSI trace consists of a channel response representation for each of the 6 resulting links, and every link includes 30 RF sub-carriers.

3.3. Machine Learning Process for Crowd Counting

For crowd counting estimation, we used standard Machine Learning classifiers so that each predicted number of people is considered as a class, so that the empty room is one class, 1 person in the room is another class and so on. Obviously in most future practical applications, classes would be numeric ranges, like 1-10 persons for one class, 11 to 50 for another one, etc.

The Machine Learning process follows the following steps:

- Step 1: Shuffle randomly the dataset rows.
- Step 2: Use feature selection criteria depending on the experiment variant.
- Step 3: Train the model with 5-folds cross-validation using the following classifiers:
    – Random Forest
    – Weighted KNN
    – Linear Discriminant
    – SVM
    – SVM with Gaussian Kernel
- Step 4: Test the model and report performance results (accuracy, AUC, etc.).

In the k-fold cross-validation process of step 3 we used a $k = 5$ instead of the more popular $k = 10$ because of the relative abundance of data, and the absence of improvements in more intensive computations resulting from increasing the k value.

3.4. Feature Extraction

A first selection of descriptor functions was made from a set of statistics commonly used in signal processing [71]. Our first objective was to test our multi-link model with all the descriptors listed in Table 2, and then to proceed to perform feature selection in order to reduce dimensionality. A more complete list of these kinds of features is provided by Di Domenico et al. [25].

Table 2. Features and descriptor functions used in the experiment.

Feature Count	Function g	Multi-Link Descriptor d
1	Mean	Mean
2	Mean	Standard deviation (SD)
3	Mean	Mean/SD
4	Mean	Spectral Energy (E)
5	Mean	Spectral Centroid (SC)
6	Mean	2nd Order Spectral Moment (SOSM)
7	Mean	2nd Order Spectral Central Moment (SOSCM)
8	Mean	Spectral Kurtosis
9–14	NOP	Mean
15–20	NOP	Standard deviation (SD)
21–26	NOP	Mean/SD
27–32	NOP	Spectral Energy (E)
33–38	NOP	Spectral Centroid (SC)
39–44	NOP	2nd Order Spectral Moment (SOSM)
45–50	NOP	2nd Order Spectral Central Moment (SOSCM)
51–56	NOP	Spectral Kurtosis

All the descriptors are relative to the magnitude of $\overline{H}_{lk}$. Notice that in the last 8 rows of Table 2 NOP refers to multi-link features without any combination function. As we have 6 links in our setup, there are 6 instances of every descriptor. It is our aim to demonstrate that this technique provides valuable information to the classification stage.

A MATLAB code was implemented to, first, obtain the Doppler spectrum from the CSI data as given by Equation (10) and then, process the feature extraction from the frequency domain function. At this time all the features where loaded into the model. The processed dataset for each of the 3 reported locations have a total of 4040 vectors of 56 features each; it also includes a class column with a labeled metadata specifying the number of people in the crowd that correspond to the row.

3.5. Feature Selection

Our experiment had four variations relative to the features taken into account, each with different criteria for feature selection, as listed below:

- Variant 1: All features-our first iteration was a 'brute force' approach to get a first estimate of the classification power of the model. The purpose of this variant is to get a "baseline" against which all other options should be compared: any subset of all features must either perform better than this one, or else perform very similarly but with less computing effort.
- Variant 2: Multicollinearity feature selection-a common approach to feature selection is to find a pair of features that are highly-correlated (i.e., above a correlation threshold) and drop one of them. We implemented this algorithm in MATLAB and applied to our dataset. The selected features using a correlation threshold of 0.85 are listed in Table 3.

Table 3. Features selected by multicollinearity method.

Feature Count	Link #	Multi-Link Descriptor d
1	Mean	Mean
2	Mean	Standard deviation (SD)
3	Mean	Spectral Energy (E)
4	Mean	Spectral Centroid (SC)
5	Mean	Spectral Kurtosis
6	1	Standard deviation (SD)
7	2	Standard deviation (SD)
8	3	Standard deviation (SD)
9	4	Standard deviation (SD)
10	5	Standard deviation (SD)
11	6	Standard deviation (SD)
12	4	Mean/SD
13	5	Spectral Energy (E)
14	1	Spectral Centroid (SC)
15	3	Spectral Centroid (SC)
16	6	Spectral Centroid (SC)
17	2	2nd Order Spectral Moment (SOSM)

- Variant 3: Mean descriptors vs multi-link descriptors-in this experiment we tested our hypothesis about the quality of multi-link descriptors (those features with multi-link descriptors without combination function). To achieve this, we compare the classification performance when only multi-link descriptors are used vs the scenario in which only multi-link mean descriptors feed into the model.
- Variant 4: Single descriptor analysis-At the opposite extreme of variant 1 we would have the use of only one feature, which is not really of practical interest, but we find it useful as another baseline. It answers the question of how well a single feature (per channel) model can perform using a multi-link approach with respect to the accuracy.

3.5.1. Variant 1: All Features

As shown in Table 4, our model delivered an outstanding accuracy performance when all the 56 selected features are used. Four out of the five classifiers were able to correctly estimate the number of people in experiment location with 100% of accuracy. Only Random forest performed just below perfect.

Table 4. Accuracy rate for model with all features used.

	All Features	
	Training Accuracy	Testing Accuracy
Random Forest	99.5%	99.7%
Weighted KNN	100.0%	100.0%
Linear Discriminant	100.0%	100.0%
SVM	100.0%	100.0%
SVM Gaussian	100.0%	100.0%

3.5.2. Variant 2: Multicollinearity Feature Selection

With the correlation criteria our set of features decreased from 56 to 17. As expected, many multi-link NOP descriptors were eliminated by the algorithm since of the strong correlation among them. However, the complete set of multi-link Standard Deviation descriptors remained. The results are shown in Table 5. All classifiers performed above 90% of accuracy.

Table 5. Accuracy rate for model with multicollinearity feature selection.

	Collinear Selection	
	Training Accuracy	Testing Accuracy
Random Forest	99.8%	100.0%
Weighted KNN	99.9%	100.0%
Linear Discriminant	92.8%	92.5%
SVM	98.9%	98.5%
SVM Gaussian	100.0%	100.0%

3.5.3. Variant 3: Mean Descriptors vs. Multi-Link Descriptors

As mentioned before, other authors have disregarded multi-link descriptors in favor of either single-link descriptors [68], or mean descriptors [25]. In this experiment, we compared both kinds of descriptors face to face. As shown in Tables 6 and 7, while using mean descriptors yield fairly good results, remarkably multi-link vectors provide perfect accuracy for all the classifiers.

Table 6. Accuracy rate for model with mean descriptors only.

	Mean Descriptors	
	Training Accuracy	Testing Accuracy
Random Forest	98.1%	98.8%
Weighted KNN	97.5%	97.6%
Linear Discriminant	86.8%	87.1%
SVM	96.5%	96.8%
SVM Gaussian	99.4%	99.7%

Table 7. Accuracy rate for model with multi-link descriptors only.

	Multi-Link Descriptors	
	Training Accuracy	Testing Accuracy
Random Forest	100.0%	100.0%
Weighted KNN	100.0%	100.0%
Linear Discriminant	100.0%	100.0%
SVM	100.0%	100.0%
SVM Gaussian	100.0%	100.0%

3.5.4. Variant 4: Single Descriptor Analysis

Now that we have empirically demonstrated that multi-link descriptors have better performance than single mean descriptors, our next step was focused on reducing the dimensionality of the model.

A hint for this task was provided by variant 2, where we applied multicollinearity feature selection. That process outlined the quality of multi-link standard deviation descriptor. As show in Table 8 multi-link SD provides 100% accuracy for the SVM Gaussian classifier.

Table 8. Accuracy rate for every individual multi-link descriptor.

Classifiers	SD	Mean	E	SC	SOSM	SOSCM	Kurtosis
Random Forest	99.7%	96.5%	98.7%	98.6%	97.8%	98.2%	98.5%
Weighted KNN	99.8%	98.5%	99.6%	99.1%	99.3%	99.2%	99.3%
Linear Discriminant	85.2%	67.6%	87.1%	77.0%	73.1%	73.1%	80.1%
SVM	90.5%	72.4%	91.8%	81.7%	79.3%	79.3%	83.6%
SVM Gaussian	100.0%	99.4%	99.9%	99.8%	99.4%	99.2%	99.8%

In order to further investigate the performance of each individual descriptor, we implemented Neighborhood Component Analysis (NCA), a multi-class, high-dimensional feature selection method initially proposed by Yang et al. [2]. This method maximizes the expected leave-one-out classification accuracy using the gradient ascent technique.

An examination of the results of NCA shown in Figure 4 indicates that: (1) Multi-link approach has higher prediction power than single-link based methods, and (2) Multi-link SD is the best single-statistic feature vector among those under review. These observations are in line with the outcomes of variants 2 and 3.

Finally, we were interested on knowing how many Wi-Fi links deliver an optimal trade-off for accuracy in our multi-link model using SD as descriptor function. To accomplish this, we ran several iterations of the model, including one additional link at each iteration and repeating the process for every possible link combination. Results in Figures 5 and 6 show that the accuracy of our model increases logarithmically with the number of available links, and this metric is near-to-perfect with a set of 6 available links (The specific numbers in this figure are too small to be read; this figure is intended to have a bird's eye view comparing the quantity of green squares (good classification) against the pink ones, as well as the way the AUC, marked in blue color, gets more and more of the area as it upper-left side grows).

3.5.5. Summary of Results by Number of Features

Table 9 shows a summary of the results from the experiment variants sorted in ascending order by the number of features for the model involved. It is worth noticing that SVM with Gaussing kernel provides perfect accuracy in all scenarios. Hence, all scenarios have at least one classifier with perfect accuracy.

Table 9. Accuracy rates for models with multi-link descriptors.

Classifiers	Multi-Link SD 6 Features	Collinear Selection 17 Features	All Multi-Link 48 Features	All Features 56 Features
Random Forest	99.7%	100.0%	100.0%	99.7%
Weighted KNN	99.8%	100.0%	100.0%	100.0%
Linear Discriminant	85.2%	92.5%	100.0%	100.0%
SVM	90.5%	98.5%	100.0%	100.0%
SVM Gaussian	100.0%	100.0%	100.0%	100.0%

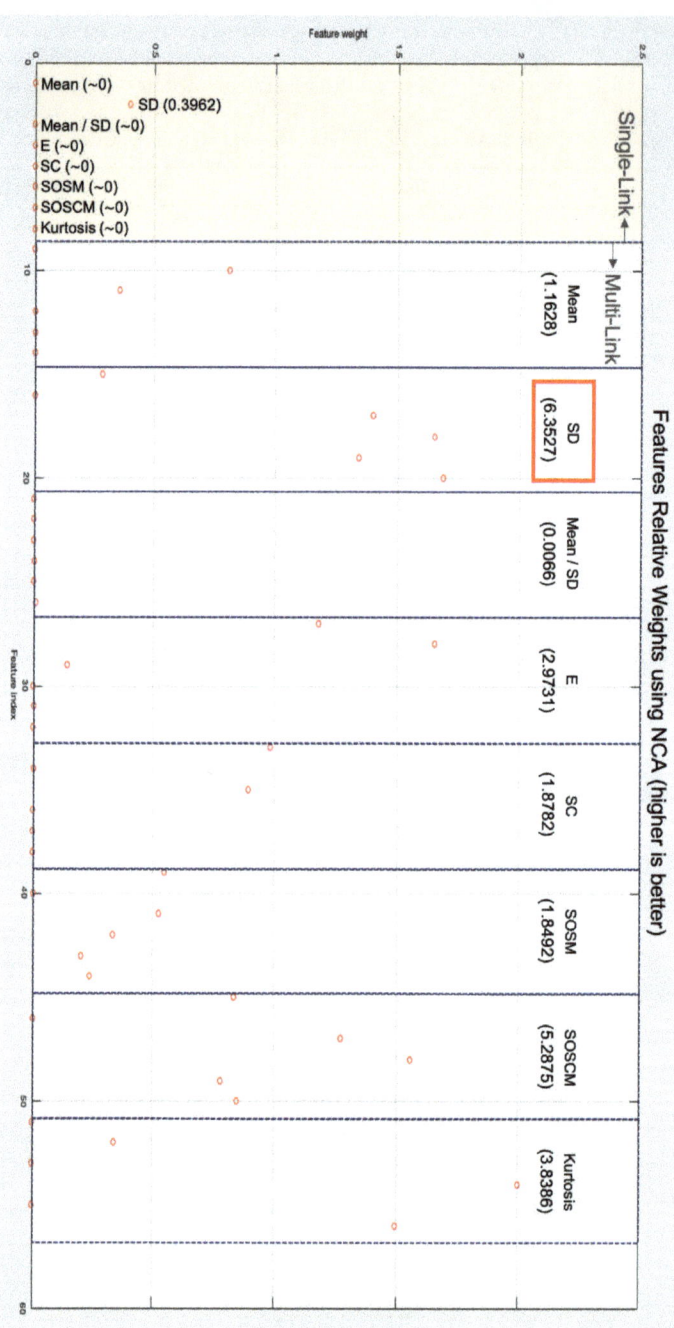

Figure 4. Relative Weights of Features using Neighborhood Component Analysis (NCA) (higher is better).

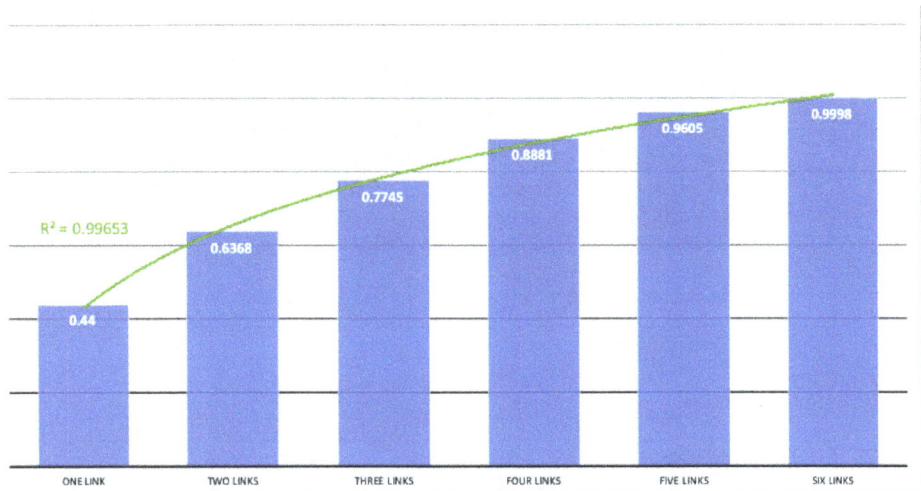

Figure 5. Accuracy of the model at incremental links.

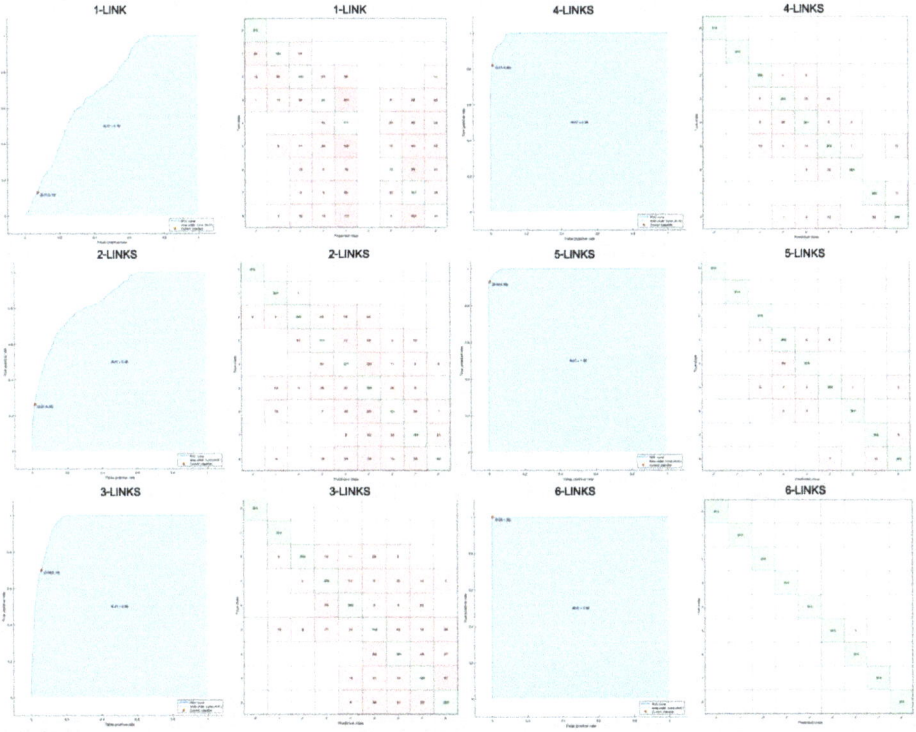

Figure 6. ROC and Confusion Matrix for incremental links.

3.5.6. Results in All Rooms

The four initial scenarios were studied using the dataset of Room A. Now, our interest was to investigate if the results in the other two rooms are similar to those we have obtained so far.

Of special interest was to validate the quality of the multi-link SD descriptor and whether the high performance of SVM-Gaussian was also observed it in the rest of the locations.

Table 10 shows the results for Room A, Room B & Room C and provides evidence that the results obtained in the first iterations with dataset of Room A extend well to the other available datasets. Multi-link SD descriptor produces accuracies of more than 97% for Random Forest, Weighted KNN and SVM-Gaussian classifiers in all rooms. We can see that for SVM-Gaussian, the model delivered perfect accuracy in Room A and nearly perfect in Room B (99.7%) and Room C (99.9%). The confusion matrices are shown in Figure 7.

Table 10. Accuracy rates for multi-link approach in all rooms.

Multi-Link SD			
	Room A	Room B	Room C
Random Forest	99.7%	97.0%	98.2%
Weighted KNN	99.8%	98.8%	99.7%
Linear Discriminant	85.2%	69.2%	64.8%
SVM	90.5%	74.2%	71.3%
SVM Gaussian	100.0%	99.7%	99.9%
Collinear Selection			
	Room A	Room B	Room C
Random Forest	100.0%	99.2%	100.0%
Weighted KNN	100.0%	100.0%	100.0%
Linear Discriminant	92.5%	87.2%	87.2%
SVM	98.5%	97.3%	99.4%
SVM Gaussian	100.0%	100.0%	100.0%
All multi-link			
	Room A	Room B	Room C
Random Forest	100.0%	100.0%	100.0%
Weighted KNN	100.0%	100.0%	100.0%
Linear Discriminant	100.0%	99.2%	98.4%
SVM	100.0%	100.0%	100.0%
SVM Gaussian	100.0%	100.0%	100.0%
All features			
	Room A	Room B	Room C
Random Forest	99.7%	99.7%	100.0%
Weighted KNN	100.0%	100.0%	100.0%
Linear Discriminant	100.0%	99.1%	98.2%
SVM	100.0%	100.0%	100.0%
SVM Gaussian	100.0%	100.0%	100.0%

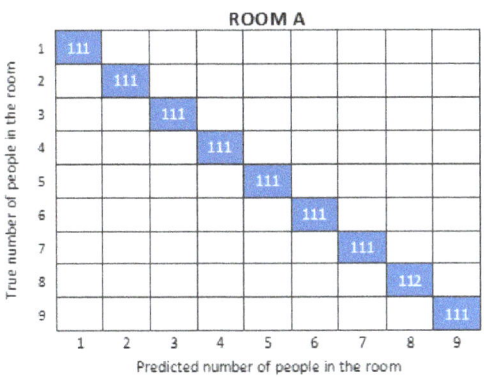
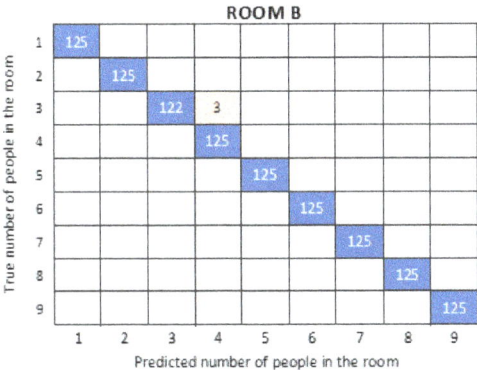
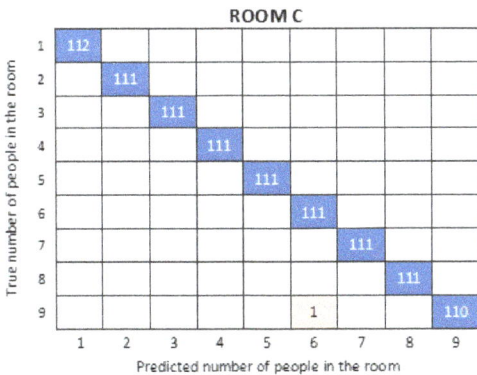

Figure 7. Confusion matrices for SVM-Gaussian with multi-link SD descriptors in all Rooms.

'All multi-link' feature set show an excellent performance since four out of five classifiers estimated the size of the crowd present in the Room with perfect accuracy. This was true for all the test cases in the available dataset.

4. Conclusions

In this paper, we have presented a novel method for crowd measuring (counting, in particular) using recognition of patterns in the Channel State Information over multiple links, and showed that the use of multiple links, instead of a single one –or the aggregation of several ones in an average– can be translated into an improved performance at least for the people counting scenario considered in the dataset we used. This was the main contribution of this work.

Using our method, based on data-driven Machine Learning supervised classifiers, we empirically demonstrated that multi-link predictors yield better performance in terms of accuracy than those that use the mean value for multi-link, or single-value of one link.

Another contributions of our work was to show that even reducing the number of features used for training and predicting with the classifiers, the performance could be maintained above that of other state-of-the-art methods. Also, we showed the prediction power of the Standard Deviation when used over the channel response data given by the Doppler Spectrum.

In Table 11, we summarize the comparison of our approach with other state-of-the-art methods.

Table 11. Benchmark of accuracy rates for state-of-the-art Wi-Fi-based crowd counting.

Author	Wi-Fi APs	Features	Classifier	Accuracy Rate	Max # of People
Di Domenico	COTS	Average Spectral Kurtosis	Naive Bayes	0.81	8
Liu	COTS	-	Deep Learning	0.82	5
Zuo	Custom	Statistics, FFT-based, Shape-based	SVM-TKL	0.96	7
Kianoush	COTS	Statistics in Space Domain	LSTM	0.99	5
Our Work	COTS	Multi-link SD in Doppler Spectrum	SVM-Gaussian	>0.99	8

For future work, we are interested in exploring the application of our method to less restricted scenarios (for instance, by increasing the maximum number of people in the crowd), and to take measurements in real-life situations. We also want to explore other crowd properties like the direction of movement and cope with limitations imposed by scale and a dynamic environment.

Author Contributions: conceptualization, R.F.B. and E.E.; investigation and methodology, E.E. and R.F.B.; supervision, R.F.B.; project administration, R.F.B.; experimental validation, E.E.; writing—original draft preparation, E.E.; writing—review and editing, R.F.B. and C.V.-R. and C.E.G.-T. and D.M.; funding acquisition, C.V.-R. All authors have read and agreed to the published version of the manuscript.

Funding: This work was supported in part by the SEP-CONACyT Research Project under Grant 256237, the School of Engineering and Sciences at Tecnologico de Monterrey and the Telecommunications Research Group. The Ph.D. studies of E.E. were supported by the CONACyT and the Tecnologico de Monterrey.

Conflicts of Interest: The authors declare no conflict of interest.

References

1. Zhang, X.; Yu, Q.; Wang, Y. Fuzzy Evaluation of Crowd Safety Based on Pedestrians' Number and Distribution Entropy. *Entropy* **2020**, *22*, 832. [CrossRef] [PubMed]
2. Yang, J.; Cheng, J.; Chen, Y. Mobile Sensing Enabled Robust Detection of Security Threats in Urban Environments. In *Lecture Notes of the Institute for Computer Sciences, Social-Informatics and Telecommunications Engineering, LNICST*; Springer: Berlin/Heidelberg, Germany, 2012; Volume 74, pp. 88–104. [CrossRef]
3. Radianti, J.; Granmo, O.C.; Bouhmala, N.; Sarshar, P.; Yazidi, A.; Gonzalez, J. Crowd models for emergency evacuation: A review targeting human-centered sensing. In Proceedings of the Annual Hawaii International Conference on System Sciences, Maui, HI, USA, 7–10 January 2013; pp. 156–165. [CrossRef]
4. Fortino, G.; Russo, W.; Savaglio, C.; Viroli, M.; Zhou, M. Modeling Opportunistic IoT Services in Open IoT Ecosystems. In Proceedings of the XVIII Workshop from Objects to Agents (WOA 2017), Scila, Italy, 2017; pp. 90–95.
5. Bouchnita, A.; Jebrane, A. A hybrid multi-scale model of COVID-19 transmission dynamics to assess the potential of non-pharmaceutical interventions. *Chaos Sol. Fractals* **2020**, *138*, 109941. [CrossRef] [PubMed]

6. Perera, C.; Zaslavsky, A.; Christen, P.; Georgakopoulos, D. Sensing as a Service Model for Smart Cities Supported by Internet of Things. *Trans. Emerg. Tel. Tech.* **2014**, *25*, 81–93. [CrossRef]
7. Luo, T.; Kanhere, S.S.; Huang, J.; Das, S.K.; Wu, F. Sustainable incentives for mobile crowdsensing: Auctions, lotteries, and trust and reputation systems. *IEEE Commun. Mag.* **2017**, *55*, 68–74. [CrossRef]
8. Joglekar, P.; Kulkarni, V. Mobile crowd sensing for urban computing. *Int. J. Latest Trends Eng. Technol.* **2017**, *7*. [CrossRef]
9. Calabrese, F.; Pereira, F.C.; Di Lorenzo, G.; Liu, L.; Ratti, C. The geography of taste: Analyzing cell-phone mobility and social events. In *International Conference on Pervasive Computing*; Springer: Berlin/Heidelberg, Germany, 2010; pp. 22–37.
10. Djenouri, D.; Laidi, R.; Djenouri, Y.; Balasingham, I. Machine Learning for Smart Building Applications: Review and Taxonomy. *Mach. Learn. Smart Build. Appl. Rev. Taxon. Acm Comput. Surv.* **2018**, *1*, 1–42. [CrossRef]
11. Bisio, I.; Lavagetto, F.; Marchese, M.; Sciarrone, A. Smartphone-centric ambient assisted living platform for patients suffering from co-morbidities monitoring. *IEEE Commun. Mag.* **2015**, *53*, 34–41. [CrossRef]
12. Wang, Y.; Yang, J.; Chen, Y.; Liu, H.; Gruteser, M.; Martin, R.P. Tracking human queues using single-point signal monitoring. In Proceedings of the 12th Annual International Conference on Mobile Systems, Applications, and Services, Bretton Woods, NH, USA, 16–19 June 2014; pp. 42–54.
13. Zhan, B.; Monekosso, D.N.; Remagnino, P.; Velastin, S.A.; Xu, L.Q. Crowd analysis: A survey. *Mach. Vis. Appl.* **2008**, *19*, 345–357. [CrossRef]
14. Wang, C.; Zhang, H.; Yang, L.; Liu, S.; Cao, X. Deep people counting in extremely dense crowds. In Proceedings of the 23rd ACM International Conference on Multimedia, Brisbane, Australia, 26–30 October 2015; pp. 1299–1302.
15. Nuño-Maganda, M.A.; Herrera-Rivas, H.; Torres-Huitzil, C.; Marisol Marin-Castro, H.; Coronado-Pérez, Y. On-Device learning of indoor location for WiFi fingerprint approach. *Sensors* **2018**, *18*, 2202.
16. Brena, R.F.; García-Vázquez, J.P.; Galván-Tejada, C.E.; Muñoz-Rodriguez, D.; Vargas-Rosales, C.; Fangmeyer, J. Evolution of Indoor Positioning Technologies: A Survey. *J. Sens.* **2017**, *2017*. [CrossRef]
17. Bishop, C.M. *Pattern Recognition and Machine Learning*; Springer: Berlin/Heidelberg, Germany, 2006.
18. Domazetovic, A.; Greenstein, L.J.; Mandayam, N.B.; Seskar, I. Estimating the Doppler spectrum of a short-range fixed wireless channel. *IEEE Commun. Lett.* **2003**, *7*, 227–229. [CrossRef]
19. Ma, Y.; Zhou, G.; Wang, S. WiFi sensing with channel state information: A survey. *ACM Comput. Surv. (CSUR)* **2019**, *52*, 1–36. [CrossRef]
20. Depatla, S.; Muralidharan, A.; Mostofi, Y. Occupancy estimation using only WiFi power measurements. *IEEE J. Sel. Areas Commun.* **2015**, *33*, 1381–1393. [CrossRef]
21. Xu, C.; Firner, B.; Moore, R.S.; Zhang, Y.; Trappe, W.; Howard, R.; Zhang, F.; An, N. SCPL: Indoor Device-Free Multi-Subject Counting and Localization Using Radio Signal Strength. In Proceedings of the 12th international conference on Information processing in sensor networks-IPSN '13, Philadelphia, PA, USA, 8–11 April 2013; ACM Press: New York, NY, USA, 2013; p. 79. [CrossRef]
22. Yuan, Y.; Zhao, J.; Qiu, C.; Xi, W. Estimating crowd density in an RF-based dynamic environment. *IEEE Sens. J.* **2013**, *13*, 3837–3845. [CrossRef]
23. Doong, S.H. Spectral human flow counting with rssi in wireless sensor networks. In Proceedings of the 2016 international conference on distributed computing in sensor systems (DCOSS), Washington, DC, USA, 26–28 May 2016; pp. 110–112.
24. Yoshida, T.; Taniguchi, Y. Estimating the number of people using existing WiFi access point in indoor environment. In Proceedings of the 6th European Conference of Computer Science (ECCS '15), Rome, Italy, 7–9 November 2015; pp. 46–53.
25. Di Domenico, S.; Pecoraro, G.; Cianca, E.; De Sanctis, M. Trained-once device-free crowd counting and occupancy estimation using WiFi: A Doppler spectrum based approach. In Proceedings of the International Conference on Wireless and Mobile Computing, Networking and Communications, Dubai, United Arab Emirates, 27–28 August 2016; pp. 1–8. [CrossRef]
26. Xi, W.; Zhao, J.; Li, X.Y.; Zhao, K.; Tang, S.; Liu, X.; Jiang, Z. Electronic frog eye: Counting crowd using WiFi. In Proceedings of the IEEE INFOCOM 2014-IEEE Conference on Computer Communications, Toronto, ON, Canada, 27 April–2 May 2014; pp. 361–369.
27. Rocklöv, J.; Sjödin, H. High population densities catalyse the spread of COVID-19. *J. Travel Med.* **2020**, *27*, taaa038. [CrossRef]
28. Yamin, M.; Ades, Y. Crowd management with RFID & wireless technologies. In Proceedings of the 1st International Conference on Networks and Communications, NetCoM, Toronto, ON, Canada, 27 April–2 May 2014; pp. 439–442. [CrossRef]
29. Wijermans, N.; Conrado, C.; van Steen, M.; Martella, C.; Li, J. A landscape of crowd-management support: An integrative approach. *Saf. Sci.* **2016**, *86*, 142–164. [CrossRef]
30. Helbing, D.; Johansson, A. Pedestrian, crowd, and evacuation dynamics. *arXiv* **2013**, arXiv:1309.1609.
31. Helbing, D.; Buzna, L.; Johansson, A.; Werner, T. Self-Organized Pedestrian Crowd Dynamics: Experiments, Simulations, and Design Solutions. *Transp. Sci.* **2005**, *39*, 1–24. [CrossRef]
32. Still, G.K. *Introduction to Crowd Science*; CRC Press: Boca Raton, FL, USA, 2014.
33. Helbing, D.; Johansson, A.; Al-Abideen, H.Z. Dynamics of crowd disasters: An empirical study. *Phys. Rev. Stat. Nonlinear Soft Matter Phys.* **2007**, *75*. [CrossRef]
34. Pathan, S.S.; Al-Hamadi, A.; Michaelis, B. Incorporating social entropy for crowd behavior detection using SVM. *Lect. Notes Comput. Sci. (Incl. Subser. Lect. Notes Artif. Intell. Lect. Notes Bioinform.)* **2010**, *6453*, 153–162. [CrossRef]
35. Vicsek, T.; Zafeiris, A. Collective motion. *Phys. Rep.* **2012**, *517*, 71–140. [CrossRef]

36. Xu, F.; Rao, Y.; Wang, Q. An unsupervised abnormal crowd behavior detection algorithm. In Proceedings of the 2017 International Conference on Security, Pattern Analysis, and Cybernetics, SPAC, Shenzhen, China, 15–17 December 2017; pp. 219–223. [CrossRef]
37. Yousefi, S.; Narui, H.; Dayal, S.; Ermon, S.; Valaee, S. A Survey on Behavior Recognition Using WiFi Channel State Information. *IEEE Commun. Mag.* **2017**, *55*, 98–104. [CrossRef]
38. Sobron, I.; Del Ser, J.; Eizmendi, I.; Velez, M. Device-Free People Counting in IoT Environments: New Insights, Results, and Open Challenges. *IEEE Internet Things J.* **2018**, *5*, 4396–4408. [CrossRef]
39. Wang, W.; Liu, A.X.; Shahzad, M.; Ling, K.; Lu, S. Device-Free Human Activity Recognition Using Commercial WiFi Devices. *IEEE J. Sel. Areas Commun.* **2017**, *35*, 1118–1131. [CrossRef]
40. Wu, C.; Yang, Z.; Zhou, Z.; Liu, X.; Liu, Y.; Cao, J. Non-invasive detection of moving and stationary human with WiFi. *IEEE J. Sel. Areas Commun.* **2015**, *33*, 2329–2342. [CrossRef]
41. Jiang, H.; Cai, C.; Ma, X.; Yang, Y.; Liu, J. Smart Home Based on WiFi Sensing: A Survey. *IEEE Access* **2018**, *6*, 13317–13325. [CrossRef]
42. Guo, L.; Wang, L.; Liu, J.; Zhou, W. A Survey on Motion Detection Using WiFi Signals. In Proceedings of the 12th International Conference on Mobile Ad-Hoc and Sensor Networks, MSN 2016, Hefei, China, 16–18 December 2016; pp. 202–206. [CrossRef]
43. Wang, X.; Yang, C.; Mao, S. PhaseBeat: Exploiting CSI phase data for vital sign monitoring with commodity WiFi devices. In Proceedings of the 2017 IEEE 37th International Conference on Distributed Computing Systems (ICDCS), Atlanta, GA, USA, 5–8 June 2017; pp. 1230–1239.
44. Han, C.; Wu, K.; Wang, Y.; Ni, L.M. WiFall: Device-free fall detection by wireless networks. *IEEE Trans. Mob. Comput.* **2014**, *16*, 271–279. [CrossRef]
45. Chowdhury, T.Z. Using Wi-Fi Channel State Information (CSI) for Human Activity Recognition and Fall Detection. Ph.D. Thesis, University of British Columbia, Vancouver, BC, USA, 2018.
46. Myrvoll, T.A.; Håkegård, J.E.; Matsui, T.; Septier, F. Counting public transport passenger using WiFi signatures of mobile devices. In Proceedings of the 2017 IEEE 20th International Conference on Intelligent Transportation Systems (ITSC), Yokohama, Japan, 16–19 October 2017; pp. 1–6.
47. Andersen, J.B.; Nielsen, J.O.; Pedersen, G.F.; Bauch, G.; Dietl, G. Doppler spectrum from moving scatterers in a random environment. *IEEE Trans. Wirel. Commun.* **2009**, *8*, 3270–3277. [CrossRef]
48. Khan, M.A.; Khan, S.F. IoT based framework for Vehicle Over-speed detection. In Proceedings of the 2018 1st International Conference on Computer Applications & Information Security (ICCAIS), Riyadh, Saudi Arabia, 4–5 April 2018; pp. 1–4. [CrossRef]
49. Depatla, S.; Mostofi, Y. Crowd counting through walls using WiFi. In Proceedings of the 2018 IEEE International Conference on Pervasive Computing and Communications (PerCom), Athens, Greece, 19–23 March 2018; pp. 1–10.
50. Shen, J.; Cao, J.; Liu, X.; Tang, S. SNOW: Detecting Shopping Groups Using WiFi. *IEEE Internet Things J.* **2018**, *5*, 1. [CrossRef]
51. Tang, X.; Xiao, B.; Li, K. Indoor Crowd Density Estimation Through Mobile Smartphone Wi-Fi Probes. In Proceedings of the IEEE Transactions on Systems, Man, and Cybernetics: Systems, Miyazaki, Japan, 7–10 October 2018; pp. 1–12. [CrossRef]
52. Depatla, S.; Mostofi, Y. Passive crowd speed estimation and head counting using WiFi. In Proceedings of the 15th Annual IEEE International Conference on Sensing, Communication, and Networking (SECON), Hong Kong, China, 11–13 June 2018; pp. 1–9.
53. Sidhu, B.; Singh, H.; Chhabra, A. Emerging Wireless Standards-WiFi, ZigBee and WiMAX. *World Acad. Sci. Eng. Technol. Int. J. Electr. Comput. Energ. Electron. Commun. Eng.* **2007**, *1*, 42–48.
54. Wilson, J.; Patwari, N. See-through walls: Motion tracking using variance-based radio tomography networks. *IEEE Trans. Mob. Comput.* **2011**, *10*, 612–621. [CrossRef]
55. Kosba, A.E.; Saeed, A.; Youssef, M. Rasid: A robust wlan device-free passive motion detection system. In Proceedings of the 2012 IEEE International Conference on Pervasive Computing and Communications, Lugano, Switzerland, 19–23 March 2012; pp. 180–189.
56. Youssef, M.; Mah, M. Challenges: Device-free Passive Localization for Wireless. *MobiCom* **2007**, 7. [CrossRef]
57. Yang, Z.; Zhou, Z.; Liu, Y. From RSSI to CSI. *ACM Comput. Surv.* **2013**, *46*, 1–32. [CrossRef]
58. Wu, K.; Xiao, J.; Yi, Y.; Chen, D.; Luo, X.; Ni, L.M. CSI-based indoor localization. *IEEE Trans. Parallel Distrib. Syst.* **2013**, *24*, 1300–1309. [CrossRef]
59. Halperin, D.; Hu, W.; Sheth, A.; Wetherall, D. Predictable 802.11 packet delivery from wireless channel measurements. *ACM SIGCOMM Comput. Commun. Rev.* **2012**, *40*, 159. [CrossRef]
60. Goldsmith, A. *Wireless Communications*; Cambridge University Press: Cambridge, UK, 2005. [CrossRef]
61. Ahmed, B.; Abdul Matin, M. *Coding for MIMO-OFDM in Future Wireless Systems*; Springer Briefs in Electrical and Computer Engineering, Springer International Publishing: Cham, Switzerland, 2015; pp. 11–21. [CrossRef]
62. Dubuc, C.; Starks, D.; Creasy, T.; Hou, Y. A MIMO-OFDM prototype for next-generation wireless WANs. *IEEE Commun. Mag.* **2004**, *42*, 82–87. [CrossRef]
63. Liu, S.; Zhao, Y.; Chen, B. WiCount: A deep learning approach for crowd counting using wifi signals. In Proceedings of the 15th IEEE International Symposium on Parallel and Distributed Processing with Applications, Guangzhou, China, 12–15 December 2017; pp. 967–974. [CrossRef]
64. Liu, S.; Zhao, Y.; Xue, F.; Chen, B.; Chen, X. DeepCount: Crowd counting with WiFi via deep learning. *arXiv* **2019**, arXiv:1903.05316
65. Cheng, Y.K.; Chang, R.Y. Device-free indoor people counting using Wi-Fi channel state information for Internet of Things. In Proceedings of the GLOBECOM 2017–2017 IEEE Global Communications Conference, Singapore, 4–8 December 2017; pp. 1–6.

66. Mabuchi, T.; Taniguchi, Y.; Shirahama, K. Person recognition using Wi-Fi channel state information in an indoor environment. In Proceedings of the 2020 IEEE International Conference on Consumer Electronics-Taiwan (ICCE-Taiwan), Taoyuan, Taiwan, 28–30 September 2020; pp. 1–2.
67. Zou, H.; Zhou, Y.; Yang, J.; Gu, W.; Xie, L.; Spanos, C. Freecount: Device-free crowd counting with commodity wifi. In Proceedings of the GLOBECOM 2017-2017 IEEE Global Communications Conference, Singapore, 4–8 December 2017; pp. 1–6.
68. Yang, D.; Wang, T.; Sun, Y.; Wu, Y. Doppler Shift Measurement Using Complex-Valued CSI of WiFi in Corridors. In Proceedings of the 2018 3rd International Conference on Computer and Communication Systems, ICCCS 2018, Nagoya, Japan, 27–30 April 2018; pp. 497–501. [CrossRef]
69. Xiao, Y. IEEE 802.11 n: Enhancements for higher throughput in wireless LANs. *IEEE Wirel. Commun.* **2005**, *12*, 82–91. [CrossRef]
70. Di Domenico, S.; De Sanctis, M.; Cianca, E.; Bianchi, G. A trained-once crowd counting method using differential wifi channel state information. In Proceedings of the 3rd International on Workshop on Physical Analytics, Singapore, 26–30 June 2016; pp. 37–42.
71. Nita, G.M.; Gary, D.E. Statistics of the Spectral Kurtosis Estimator. *Publ. Astron. Soc. Pac.* **2010**, *122*, 595–607. [CrossRef]

Constructing Emotional Machines: A Case of a Smartphone-Based Emotion System

Hao-Chiang Koong Lin [1], Yu-Chun Ma [1,*] and Min Lee [2]

[1] Department of Information and Learning Technology, National University of Tainan, Tainan 700, Taiwan; koong@gm2.nutn.edu.tw
[2] Zero Dimension Technology Co., Ltd., Taichung 407, Taiwan; axdrolee@gmail.com
* Correspondence: d10455003@stumail.nutn.edu.tw; Tel.: +886-6-213-3111

Abstract: In this study, an emotion system was developed and installed on smartphones to enable them to exhibit emotions. The objective of this study was to explore factors that developers should focus on when developing emotional machines. This study also examined user attitudes and emotions toward emotional messages sent by machines and the effects of emotion systems on user behavior. According to the results of this study, the degree of attention paid to emotional messages determines the quality of the emotion system, and an emotion system triggers certain behaviors in users. This study recruited 124 individuals with more than one year of smartphone use experience. The experiment lasted for two weeks, during which time participants were allowed to operate the system freely and interact with the system agent. The majority of the participants took interest in emotional messages, were influenced by emotional messages and were convinced that the developed system enabled their smartphone to exhibit emotions. The smartphones generated 11,264 crucial notifications in total, among which 76% were viewed by the participants and 68.1% enabled the participants to resolve unfavorable smartphone conditions in a timely manner and allowed the system agent to provide users with positive emotional feedback.

Keywords: effective computing; emotion system; emotional machine; agent; human–machine interface

1. Introduction

Human interactions have gradually evolved toward diverse methods of interaction beyond conventional in-person ones. Software has become integral to human life, and software development has led to numerous advances. Currently, human–machine interactions are more prevalent than person–person interactions. The human–machine interface plays a crucial role in the diversified interactions between humans and machines, especially in the era of the Internet of Things (IoT), by enabling information exchange between humans and machines [1]. With a high market penetration rate, the smartphone market has matured. The high rates of smartphone ownership indicate that smartphone use has become widespread in daily life [2]. Since 2015, the percentage of smartphone users has continued to increase across multiple countries and age groups [3]. In the IoT era, wearable technology is in a rapid growth phase and has attracted increasing attention from both industry and academia over the past decade [4]. The use of wearable devices and IoT services in people's daily lives is increasing, and individuals are exposed to diverse software and hardware services. An important objective of human–machine interaction, especially in the field of machine emotion expression, is to make the behavior of a machine more similar to that of a human [5].

In an ideal intelligent interactive environment, a machine has the same external stimulus perception ability as a human does. Such a machine can conduct a simulation to recognize, process, and understand external stimuli, and has the ability of emotion computation [5]. Currently, there is no single, well-developed human–machine interaction mechanism that can satisfy the requirements of those working with various related machines. This phenomenon is a key challenge in the field of human–machine interfaces. Human emotion is a wide research topic that covers different fields, including psychology, neuroscience, health science, and engineering. An accurate model of human emotion would be beneficial for developing an emotion recognition system and its applications [6]. The concept of an emotion system has been applied in various fields and has yielded many benefits. Emotions can be recognized from speech, activity, and facial expressions. For example, appropriate and personalized learning can be achieved by applying the affective tutoring system (ATS) in teaching through emotional computing technology. Thus, each student faces their own emotional learning system on their smartphones for learning [7]. The system will analyze the emotions of the student at each moment and complete the course content according to the student's learning status [8]. However, emotion systems are mostly designed to serve a specific system. Because few studies have investigated attitudes toward machines, an emotion system was designed in this study to assign emotions to machines, thus transforming cold, impersonal machines into life-like entities, and this study explored users' attitudes toward emotional machines. When machines have emotions, the positive effect is that machines with emotional intelligence are not the same as our "human" friends. The former will know everything about us, and we cannot hide anything from them. They may know us a lot more than ourselves. Therefore, they may be able to help us make better decisions and choices in life. It is invaluable to have such a "person" to accompany us in our lives. Making artificial intelligence more humane and empathetic is the future development goal of artificial intelligence [9]. Emotion is a necessary condition for communication. The understanding of both parties can be improved when there is more human–computer interaction. The emotions of the machine can improve human emotions in daily life situations [10]. If a machine has emotions, it will not only reduce the gap between the machine and the user, but also trigger or change the behavior of the user, allowing the user to actually adopt specific behaviors, thereby creating commercial value [11].

In this study, an emotion system was developed and installed on smartphones to enable them to exhibit emotions. The emotion system (in the form of an app) was designed for Android smartphones to track. Users have access to information on the past and present conditions of their smartphones. More warm and humane services can be realized by enabling emotions in machines. Thus, friendly artificial intelligence can be realized. The test subject will need a smartphone with the Android operating system. There will be compatibility issues when designing Android applications. When using the same application on different models, the application system cannot be used normally in some circumstances. Moreover, the app is equipped with a system agent that discloses smartphone conditions in a human-like manner. When a smartphone has an unfavorable condition (e.g., being extremely low on battery), the system automatically informs the user by sending them a timely notification. In addition, the system agent within the application notifies users of any smartphone problems. The agent also analyzes the view rate of such notifications and the instruction execution rate, which allows program designers to determine the effect of the developed emotion system on user behavior. After smartphone malfunctions are resolved, the agent provides users with positive emotional feedback. This study analyzed the following aspects:

(1). Factors requiring attention during the development of emotional machines.
(2). Users' attitudes toward emotional messages.
(3). Effects of the developed emotion system on user behavior.

2. Literature Review
2.1. Affective Computing

The concept of a human–machine interface was proposed by Norman, and Picard [12] laid the foundation for affective computing. Following these foundational works, the application of affective computing in human–machine interfaces received considerable attention [13]. Affective computing is an interdisciplinary field that focuses on computer models and methods for recognizing and expressing emotions [14]. Affective computing was originally proposed in 1997 by Rosalind Picard from the MIT Media Lab [12]. It results from biomedical engineering, psychology, and artificial intelligence. Affective computing aims to allow computer systems to detect, use, and express emotions [15]. It is a constructive and practical approach that focuses on improving human-like decision support and human–machine interaction [16].

In the field of human–machine interfaces, user experience is essential, and users' emotions and reactions affect user satisfaction [17]. Many studies on emotion systems have established affective tutoring systems [18] and verified the positive influence of emotion system interfaces on learning. In addition, affective computing has been incorporated into human–machine interfaces. For example, the eMoto system proposed by Sundström et al. in [19] is an emotional text messaging interface that builds on the physiological data (e.g., body movement data) captured by a smart pen to generate graphical and expressive backdrops for messages. The Affective Diary system proposed by Ståhl et al. in [20] collects data on user emotions through a physiological sensor on the day of use. The results, represent a user's affective memories of the day of use. In summary, the aforementioned research indicates that affective computing positively affects human–machine interfaces when emotion systems are appropriately designed [21].

2.2. Emotion Systems

Research on affective computing can be divided into two main branches that focus on (1) detecting and recognizing emotional messages, and (2) expressing emotions. This study focused on the expression of emotions. Bretan et al. in [22] constructed a robot with the ability to express emotions and process languages. They found that participants who interacted with the robot exhibited a greater sense of participation and joy than those who did not. This result may be attributed to users finding a system more valuable when it provides emotional feedback. Therefore, the effective expression of emotions is key to establishing an appropriate emotion system. Research has also been conducted on enabling machines to exhibit emotions. For example, Bates conducted a preliminary study by developing a simplified emotional agent that expresses fundamental emotional states [23]. Subsequently, Ushida et al. in [24] developed a set of emotion modules in which emotions are expressed through a life-like emotional agent. Maria and Zitar [25] modeled artificial emotions through agents and proposed emotional algorithms for the operation of the emotion module. Evidently, the literature on incorporating emotions into machines is gradually expanding. All studies on this topic have indicated that emotion systems must be equipped with an emotion module that satisfies the research objective.

2.3. Emotional Expression

Emotional expression refers to how emotions are conveyed. Emotion systems are generally equipped with emotion modules and use emotional expression as the framework to support the operational processes of emotion modules. Emotional expression research is based on two mainstream theories: discrete emotion theory and continuous emotion theory. Discrete emotion theory is characterized by a discrete classification of emotions. The most prominent discrete categorization is that proposed by Ekman [26] and comprises fear, anger, disgust, sadness, happiness, and surprise [27].

Proponents of continuous emotion theory argue that emotions can be fully expressed through neural and physiological systems. In discrete emotion theory, emotions are classified according to the neurophysiological systems associated with them. Continuous emotion theory was first proposed in 1897 by the psychologist Wundt, who divided emotions into three dimensions: pleasurable versus unpleasurable, arousing versus subduing, and straining versus relaxing [28]. Subsequently, as an extension of Wundt's theory, Woodworth and Schlosbeg [29] reformulated the three dimensions of emotions as pleasantness–unpleasantness, attention–rejection, and level of activation. In addition to these three-dimensional models, two-dimensional models, including the circumplex model [30], vector model [31], and PANA model [32], have been widely applied in the literature. However, the present study adopted continuous emotion theory and computed emotional expression.

3. Research Methods

3.1. Assigning Emotions to Smartphones

To assign emotions to smartphones, this study first designed emotion modules in accordance with Picard's four motivations for enabling machines to exhibit emotion. Then, the circumplex model was adopted as the framework for the expression of emotions. The emotion modules were designed to perform affective computing in a two-dimensional emotional space (Figure 1).

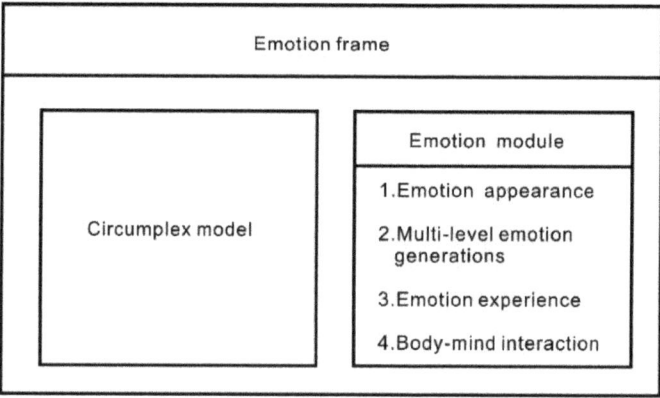

Figure 1. Emotional framework of the designed system.

3.1.1. Developing Emotional Expression

Russell represented emotional expression by using a spatial model in which affective concepts fall in a circle in the following order: pleasure, 0°; excitement, 45°; arousal, 90°; distress, 135°; displeasure, 180°; depression, 225°; sleepiness, 270°; and relaxation, 315° [20]. The present study employed Russell's circumplex model of affect as the foundation for constructing a two-dimensional integer space containing valence and arousal dimensions, which are represented on the horizontal and vertical axis, respectively. The center of the space represents the origin (0, 0). According to an analysis of system requirements, a clear correspondence between emotions and behaviors should be achieved. To satisfy this requirement, the system must convert quantitative emotional data into categories of emotions. In the aforementioned emotion coordinates, each quadrant covers 45°; thus, the two-dimensional space is divided into eight categories of emotions. The first to fourth quadrants represent happiness/joy, anger/dissatisfaction, sadness/pain, and calmness/peace, respectively (Figure 2).

Figure 2. Emotion categories in the vector space.

3.1.2. Developing Emotional Representation

Machines should exhibit emotions in a unique manner, and different emotions for the same behavior should have different manifestations (Figure 3).

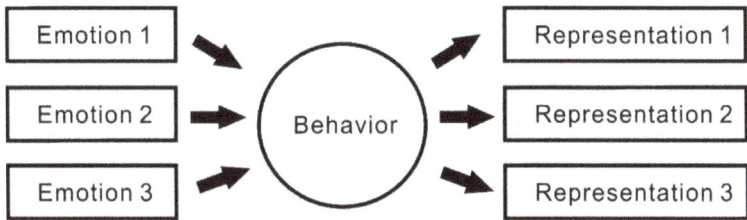

Figure 3. Schematic of emotional representation.

The agent built within the system can express the emotional state and activities of the system. The agent follows a certain schedule and plans specific activities for each time period. An agent's engagement in certain activities determines the emotional state of the system, projects it to the activities in practice, and results in the generation of emotional representations (Figure 4).

Figure 4. Representations of happiness (**left**) and pain (**right**) during eating.

According to the principle of emotion representation, different emotions may correspond to the same behavior, producing different representations. Each activity corresponds to eight types of emotions. Thus, each activity will have eight emotion representations. There are 12 types of activity items and 8 types of emotions in this system. Based on the principle of emotion representation, the system produces 96 emotion representations. At each specific time, the agent has a type of activity to be performed. However, if the same activity is performed throughout the day, the activity of the agent is the same throughout the day for the user, resulting in no variability. The system sets the activity category as an activity collection, and each activity collection contains an activity subcollection. For example, let us consider the activity collection of eating. Its subcollection will include eating French fries, eating burgers, eating cake, etc. Therefore, users will observe different agent activities at the same time on different days.

3.1.3. Principles for Multidimensional Emotion Generation

According to the concept of multidimensional emotion generation, emotions can be generated through various methods. Multidimensional emotions were generated in this study through the adoption of the reason-generated operation and the quick and dirty method proposed by Picard [12]. Specifically, this study's system operates as follows. First, reason-generated events trigger Picard's algorithm. The system then processes the emotional parameters of the emotional event table and determines the system's present emotional state through the circumplex model. Picard's algorithm is subsequently executed to obtain the new emotional parameters of the events. The emotional event table is then updated with the new parameters in accordance with the principle of reason-generated operation (Figure 5).

Let $P(i,j), E(m,n) \in$ Circumplex model

Case 1:

$i * m < 0$

$\Rightarrow E(m,n) * c = E(cm,cn)$ where $c > 0$

Case 2:

$i * m > 0 \cap |i| > 0.75$ valence-axis

Figure 5. Principle of reason-generated operation.

The quick and dirty method was implemented according to the emotional event table with timely responses (Table 1). Quick and dirty events prompted the system to generate emotional representations rapidly without updating the emotional event table (Figure 6).

Table 1. List of quick and dirty events.

Event Content	Positive and Negative Emotions	Intensity of Emotion
Power < 25%	−4	−4
RAM dosage > 75%	−4	−4
Call notification	0	1
Charging	1	2
When the power < 50%, sleeping, and charging	2	3

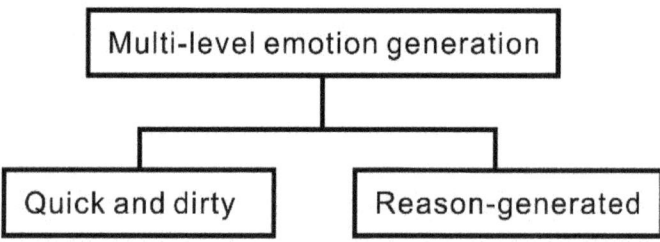

Figure 6. Schematic of multidimensional emotion generation.

3.1.4. Developing Emotional Experiences

Emotional experience implies that a machine can recognize an event it had encountered and knows which emotional reactions that type of event typically elicits. The proposed system is a practical cognitive module. First, it establishes an emotional event table to classify an unfavorable smartphone event and record the emotional parameters induced by this event. The combination of emotional parameters induced by an unfavorable event is termed an emotional event table (Table 2).

Table 2. Emotional event table.

Event Id	Name	Valence	Arousal	Quick and Dirty
1	Battery > 75	3	2	false
2	Battery 50~75	1	1	false
3	Battery 25~50	−1	−2	false
4	Battery < 50	−4	−4	true
5	RAM > 75	1	−4	true
6	RAM 50~75	−1	2	false
7	RAM 25~50	1	1	false
8	RAM < 50	3	2	false
9	Storage > 90	−4	−4	false
10	Storage 70~90	−1	−2	false
11	Storage 40~70	1	1	false
12	Storage < 40	3	2	false
13	Network increased	−1	−1	false
14	Network decreased	1	1	false
15	Temperature good	1	1	false
16	Overheat	−1	−1	true
17	Charging	1	2	true
18	Call incoming	0	1	true
19	Sleep-negative charging	2	3	true

An emotional event stores the parameters of state events, and emotional events must be mapped to a two-dimensional emotional space. The lengths of the horizontal and vertical axes are related to the absolute values of the parameters of all events constituting an emotional event. According to the initial parameters of the emotional event in Table 2, the horizontal axis length (valence) of the two-dimensional emotional space in the developed system is 31, and the vertical axis length (arousal) is 37. The system event is divided into categories such as battery status, network status, memory status, storage space status, and incoming call notification. Four battery statuses exist, namely: above 75%, 50–75%, 25–50%, and below 25%. A battery level of 25% or lower is a quick and dirty event. Four memory statuses exist: above 75%, 50–75%, 25–50%, and below 25% memory use. Memory use of above 75% is a quick and dirty event. Four storage space statuses exist, namely: above 90%, 70–90%, 40–70%, and below 40% storage use. The memory card has no direct influence on the smartphone status, thus no quick and dirty events related to storage are handled by the system. The network status event category has two subcategories, namely increased and decreased network traffic. The system regularly calculates the network traffic and records the average traffic during different periods. When the average flow rate in an interval increases, the system triggers an event that increases the flow rate. When the average flow rate in an interval decreases, the system triggers a flow decrease event. The call notification setting is also set as an emotional event. When the smartphone receives an incoming call, the system triggers an emotional event. This emotional event has no parameters for positive and negative emotions; it only has emotional intensity parameters.

Whenever events that are unfavorable to the smartphone occur, the system identifies the corresponding emotional event from the emotional event table and obtains the corresponding emotional parameter. The system may also update the emotional event with new emotional parameters according to the principle of reason-generated operation.

3.1.5. Developing Psychophysical Interactionism

Psychophysical interactionism refers to the interaction between the software and hardware conditions of a machine. The emotions of a machine are affected when both the aforementioned factors are considered. For example, sufficient memory space increases efficiency by providing sufficient time for computation, enables software programs to run smoothly, and ensures that the smartphone remains in a positive emotional state. The emotional event table records software and hardware events. When predicting the system's emotional responses to specific events, the hardware condition is considered. For example, when a smartphone is completely charged, the system must permit higher memory use to prevent negative emotions arising from increased memory use.

3.1.6. Method for Detecting Smartphone Conditions

In response to changes in a smartphone's condition, the developed system identifies the emotional event corresponding to the changes and the corresponding emotional parameters from the emotional event table. The developed system is a program that runs in the background and remains visible on the home screen. Therefore, when a user turns on their mobile smartphone, the developed system runs automatically. A priority table comprising emotional events to be prioritized in selection is established by the system. This system and the algorithm serve as the guide for selecting the emotional events triggered by changes in smartphone condition. The system detects the condition of the smartphone through the native battery life tracking and notification functionality of Android systems. Each event detection is associated with a given level of battery.

To establish an event selection mechanism, the developed system divides the battery charge level into three intervals: (1) >75%, (2) 35–75%, and (3) <35%. The events prioritized and selected by the system vary with the battery charge interval. Specifically, if a smartphone falls within the first and second battery intervals, the system prioritizes events with positive and negative emotions, respectively. Under these intervals, when more than one event occurs, the system prioritizes notifications regarding limited storage space, followed by increased network traffic, high smartphone temperature, excessive random-access memory (RAM) usage, and low battery. If none of the aforementioned events occur, the system randomly selects other suitable events that match the condition of the smartphone. When the charge level falls in the third interval, the system prioritizes events with negative emotions. For this interval, when more than one event occurs, the system prioritizes notifications related to increased network traffic, followed by those related to high smartphone temperature, excessive RAM usage, insufficient storage space, and low battery. Because negative events are necessarily highlighted in the third interval, no other emotional events are randomly selected.

Quick and dirty events are prioritized in the selection of unfavorable smartphone events. The quick and dirty events are as follows: battery at <25% charge, >75% of RAM used, overheating battery, battery being charged, an incoming call, and the system having a negative emotional state when the smartphone is being charged in sleep mode. If two events occur simultaneously, the developed system prioritizes notifications in the following order: incoming call, the system having a negative emotional state when the smartphone is being charged in sleep mode, battery being charged, battery at <25% charge, overheating battery, and >75% of RAM used. Such prioritization ensures event uniqueness when multiple quick and dirty events occur simultaneously.

3.2. Assigning Emotions to Messages

According to the function of the developed system, a message pertaining to the smartphone's condition is generated by the system. For example, when a large amount of memory is used, the system provides a corresponding notification to the user in the form of a text message containing emotional phrases. The types of emotions in text messages are determined by the system's present emotion (Figure 7).

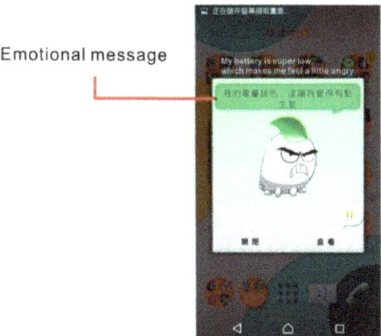

Figure 7. An emotional message generated by the system.

After a user clicks on the system agent, the agent provides a few sentences of feedback that comprises two parts. The first part describes the present condition of the smartphone, and the second part describes the system's current emotional state (Figure 8).

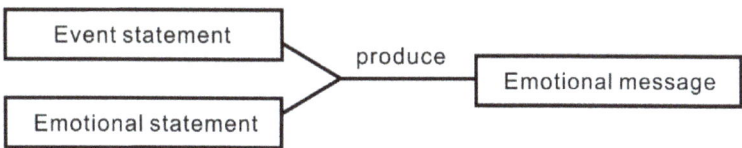

Figure 8. Schematic of the system message generation method.

The messages sent by the developed system depend on the events that occur and the system's emotions. Sentences about such events contain emotional parameters, and events are divided into positive and negative events according to the system's emotions. Sentences about emotions are, by nature, positive or negative. Therefore, the following four combinations are produced through the combination of sentences about events and emotions: positive–positive, positive–negative, negative–positive, and negative–negative. The system redefines these four combinations into two categories: consistent and inconsistent. Consistency between sentences about an event and emotions suggests that the event and emotional state elicit the same emotions (positive or negative).

3.3. Developing a Crucial Notification Mechanism

On the basis of the push notification function native to the Android operating system, a crucial notification mechanism was designed for the developed system. The developed system uses push notifications to highlight events with negative emotional parameters (Figure 9).

Figure 9. Crucial notification mechanism.

The system's notification mechanism was designed to determine whether users are prompted by crucial notifications to follow related instructions and complete a specific action. A schematic of the system's software architecture is shown in Figure 10, in which four modules are displayed.

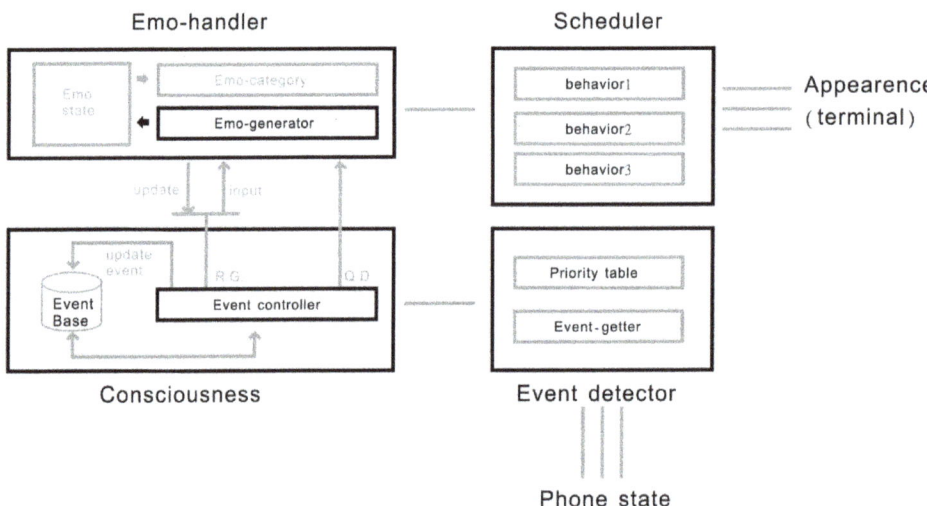

Figure 10. Schematic of the software architecture.

The event detector is a module for detecting the smartphone's condition. The event getter in the aforementioned module, which is driven by the operating system, determines the current state of the smartphone according to the order specified in the priority table to define the smartphone's condition. The event getter can notify users of any changes in the smartphone's condition by sending crucial notifications. Once users have read these notifications, the system tracks their follow-up actions, that is, whether they follow the instructions and complete a specific action.

The event detector transfers the smartphone's condition to a cognitive module of emotions called "consciousness." The consciousness module matches the smartphone condition with an event in the emotional event table. Conditions identified as quick and dirty events are then forwarded to the Emo-handler module, whereas those identified as reason-generated events have their parameters reset according to the principle of affective computing. The emotional event table is then updated with new emotional parameters, which is then forwarded to the emo-handler module.

On the basis of James Russell's emotional expression method, the emo-handler module executes events and affective computing. After receiving information regarding an event from the consciousness module, the emo-generator projects the event onto a two-dimensional emotion space to determine the present emotions of the system; thus, the emo-generator can calculate the emotional parameters of the event and the system emotions to update the emotional state of the system. Subsequently, the event is recorded as a past event in the archive. In addition, the emo-handler module transmits the system's emotional state to the scheduler module, which converts emotional parameters into categories of emotions. This conversion allows the scheduler module to generate emotional representations effectively. After obtaining the system's emotions, the scheduler module determines the agent's current activity table and generates corresponding representations according to the principle of emotional representation.

3.4. Research Tools

3.4.1. System Usability Scale

In this study, the System Usability Scale (SUS) developed by Digital Equipment Co Ltd. in 1986 was used to assess users' evaluation of the system usability. This scale contains 10 items, and each item is scored using a 5-point Likert scale ranging from 1 to 5. A higher score indicates higher user satisfaction with the system's usability. The SUS, which is presented in Table A1, is reliable, fast, convenient, and inexpensive [33].

3.4.2. Research Questionnaire

The four dimensions of the research questionnaire are based on the architecture proposed by Picard [34]. The research questionnaire is divided into four parts to (1) test whether users believe that the machine has emotions, and determine to (2) users' attention to emotional information, (3) users' interest in emotional information, and (4) the effect of emotional information about users. This questionnaire is answered using a 5-point Likert scale ranging from 1 to 5. In this study, expert interviews were conducted in three rounds by using the Delphi method [35]. The experts had expertise in different fields, such as affective computing, interaction design, and user experience. Through factor analysis, the researchers selected 18–23 items for which consensus was reached among the experts. The research questionnaire is presented in Table A2.

4. Research Results and Analysis

4.1. Reliability Analysis

This study involved 124 individuals as participants, among which there were 62 males and 62 females. These individuals were college students between the ages of 20 and 25 years, all of whom had experience in using smartphone applications and had been using smartphones for at least one year. Among the participants, 32 had knowledge about affective computing, whereas the remaining participants had no experience with affective computing systems. The experiment lasted for 2 weeks. The research questionnaire's reliability was evaluated using Cronbach's α. The Cronbach's α value ranges between 0 and 1. The criteria for identifying the internal consistency of a questionnaire are presented in Table 3. The questionnaire used in this study had excellent reliability ($\alpha = 0.914$, Table 4). The reliability of the research questionnaire for different aspects is presented in Table 5. The questionnaire had high reliability for all the considered aspects.

Table 3. Ranges of Cronbach's alpha for different categories of internal consistency.

Cronbach's Alpha	Internal Consistency
$\alpha \geq 0.9$	Excellent
$0.9 > \alpha \geq 0.8$	Good
$0.8 > \alpha \geq 0.7$	Acceptable
$0.7 > \alpha \geq 0.6$	Questionable
$0.6 > \alpha \geq 0.5$	Poor
$0.5 > \alpha$	Unacceptable

Table 4. Overall reliability of the research questionnaire.

Cronbach's Alpha	Internal Consistency
0.914	Excellent

Table 5. Reliability of different aspects of the research questionnaire.

Research Orientation	Cronbach's Alpha	Internal Consistency
Existence of Emotions in Machines	0.790	Acceptable
Attention Paid to Emotional Messages	0.792	Acceptable
Interest Toward Emotional Messages	0.707	Acceptable
Effects of Emotional Messages	0.751	Acceptable
System Usability	0.807	Good

4.2. Descriptive Statistics

4.2.1. Existence of Emotions in Machines

For items regarding participant perceptions of whether the machine had emotions, the overall mean score provided by the participants was 3.6, with a standard deviation (SD) of 0.6 and a standard error (SE) of 0.1. A total of 76 participants provided above-average scores, and 48 participants provided below-average scores (Table 6).

Table 6. Descriptive statistics for items regarding the existence of emotions in machines.

Mean	Max	Min	SD	SE
3.6	5	2	0.6	0.1

The mean scores for all the items regarding the existence of emotions in machines were >3 points. The frequency distribution results for the existence of emotions in machines are presented in Table 7. The modal scores of Q1 and Q3 was "agree", and the Q2 score was "neutral," which indicates that most of the participants perceived their machines to have emotions.

Table 7. Frequency distribution table for items regarding the existence of emotions in machines.

Item	Strongly Disagree	Disagree	Neutral	Agree	Strongly Agree
Q1	0	0	32 (25%)	68 (54%)	24 (19%)
Q2	0	20 (16%)	72 (58%)	24 (19%)	8 (6%)
Q3	0	8 (6%)	40 (32%)	60 (48%)	16 (12%)

4.2.2. Attention Paid to Emotional Messages

For items regarding the attention paid to emotional messages, the participants provided an overall mean score of 3.0, with an SD of 0.8 and an SE of 0.1. A total of 72 participants had above-average scores, and 52 participants had below-average scores (Table 8).

Table 8. Descriptive statistics for items regarding the attention paid to emotional messages.

Mean	Max	Min	SD	SE
3	5	1	0.8	0.1

The mean scores of all the items regarding the attention paid to emotional messages were >3 points. Table 9 presents the frequency distribution results for items regarding the attention paid to emotional messages. The modal score of Q6 and Q8 was "agree"; that of Q4, Q5, and Q7 was "neutral"; and that of Q9 was "strongly disagree." This is not a positive result on a 5-point Likert scale; thus, on average, the participants did not pay attention to the emotional information provided by the system. The participants paid attention to emotional messages when the agent was in sight. Although the participants were willing to read the emotional messages produced by the system, they tended not to open the system when the agent was not in sight. Some participants voluntarily turned on their smartphones to check the system's emotions.

Table 9. Frequency distribution table for items regarding the attention paid to emotional messages.

Item	Strongly Disagree	Disagree	Neutral	Agree	Strongly Agree
Q4	0	36 (29%)	40 (32%)	28 (22%)	20 (16%)
Q5	8 (6%)	24 (19%)	40 (32%)	32 (25%)	20 (16%)
Q6	8 (6%)	12 (9%)	28 (22%)	48 (38%)	28 (22%)
Q7	28 (22%)	28 (22%)	40 (32%)	20 (16%)	8 (6%)
Q8	4 (3%)	16 (12%)	48 (38%)	48 (38%)	8 (6%)
Q9	48 (38%)	28 (22%)	28 (22%)	16 (12%)	4 (3%)

4.2.3. Interest in Emotional Messages

For items regarding user interest in emotional messages, the participants had an overall mean score of 3.5, with an SD of 0.8 and an SE of 0.2. A total of 80 participants had above-average scores, and 44 participants had below-average scores (Table 10).

Table 10. Descriptive statistics for items regarding user interest in emotional messages.

Mean	Max	Min	SD	SE
3.5	5	1	0.8	0.2

The mean scores of all the items regarding user interest in emotional messages were >3 points. The frequency distribution results for items regarding user interest in emotional messages are presented in Table 11. The modal score of Q10, Q11, and Q12 is "agree." For Q11, 54% of the participants agreed that emotional information is interesting, which is a positive result on a 5-point Likert scale. This result indicates that participants were generally interested in the emotional message compiled by the system.

Table 11. Frequency distribution table for items regarding user interest in emotional messages.

Item	Strongly Disagree	Disagree	Neutral	Agree	Strongly Agree
Q10	12 (9%)	16 (12%)	24 (19%)	56 (45%)	16 (12%)
Q11	4 (3%)	12 (9%)	40 (32%)	68 (54%)	0
Q12	8 (6%)	12 (9%)	28 (22%)	48 (38%)	28 (22%)

4.2.4. Effects of Emotional Messages

For items regarding the effects of emotional messages, the participants had an overall mean score of 3.5, with an SD of 0.7 and an SE of 0.1. A total of 72 participants had above-average scores, and 52 participants had below-average scores (Table 12).

Table 12. Descriptive statistics for items regarding the effects of emotional messages.

Mean	Max	Min	SD	SE
3.5	5	1	0.7	0.1

The mean scores of all items regarding the effects of emotional messages were >3 points. The frequency distribution results for the aforementioned items are presented in Table 13. The modal score of Q13, Q14, Q15, Q16, and Q18 was "agree," whereas that of Q17 was "disagree," which represents a positive result on a 5-point Likert scale. The aforementioned result indicates that emotional messages affected most users and prompted them to reexamine and be mindful of their smartphone usage habits. Moreover, participants who read the emotional messages, which served as a trigger for reflections, reflected on their smartphone usage habits.

Table 13. Frequency distribution table for items regarding the effects of emotional messages.

Item	Strongly Disagree	Disagree	Neutral	Agree	Strongly Agree
Q13	4 (3%)	12 (9%)	16 (12%)	56 (45%)	36 (29%)
Q14	0	16 (12%)	32 (25%)	56 (45%)	20 (16%)
Q15	4 (3%)	16 (2%)	8 (6%)	60 (48%)	36 (29%)
Q16	4 (3%)	8 (6%)	36 (29%)	60 (48%)	16 (12%)
Q17	4 (3%)	40 (32%)	32 (25%)	40 (32%)	8 (6%)
Q18	12 (9%)	28 (22%)	24 (19%)	44 (35%)	16 (12%)

4.2.5. System Usability

For items regarding system usability, the participants had an overall mean score of 4.1, with an SD of 0.5 and an SE of 0.1. A total of 60 participants had above-average scores, and 64 participants had below-average scores (Table 14).

Table 14. Descriptive statistics for items regarding system usability.

Mean	Max	Min	SD	SE
4.1	5	2	0.5	0.1

The total score for the aforementioned items was 78.1 points, which indicated that most of the participants found the developed system to have acceptable usability and be unobtrusive in terms of daily smartphone use (Table 15).

Table 15. Frequency distribution table for items regarding system usability.

Item	Strongly Disagree	Disagree	Neutral	Agree	Strongly Agree
Q1	0	4 (3%)	48 (38%)	56 (45%)	16 (12%)
Q2	0	0	24 (19%)	68 (54%)	32 (25%)
Q3	0	0	20 (16%)	68 (54%)	36 (29%)
Q4	0	0	16 (12%)	40 (32%)	68 (54%)
Q5	0	8 (6%)	44 (35%)	48 (38%)	24 (19%)
Q6	0	4 (3%)	32 (25%)	48 (38%)	40 (32%)
Q7	0	0	16 (12%)	52 (41%)	56 (45%)
Q8	0	4 (3%)	28 (22%)	36 (29%)	56 (45%)
Q9	0	0	12 (9%)	52 (41%)	60 (48%)
Q10	0	0	16 (12%)	48 (38%)	60 (48%)

4.2.6. View Rate of Crucial Notifications and Instruction Execution Rate

Information on crucial notifications was obtained from the operation data collected by the developed system. During the experiment, 11,264 crucial notifications were generated, of which 8636 were viewed by the participants. The overall view rate of crucial notifications was 76.6%, and 7672 instructions were executed to resolve unfavorable smartphone conditions and restore the device to normal conditions. The overall instruction execution rate was 68.1%. In this study, a high view rate of the system's crucial notifications and a high instruction execution rate were achieved.

4.3. Experimental Results

4.3.1. Factors Influencing the Participants' Perception of Emotions in Machines

The participants provided a mean score of 3.0 for items regarding the attention paid to emotional messages. However, this study revealed that the participants rarely paid active attention to emotional messages and were sometimes indifferent toward these messages. Thus, the developed system's emotional messages can be improved. The participants provided a mean score of 3.6 for items related to their perception regarding whether their machine had emotion. Generally, those who devote much attention to emotional messages tend to perceive that their machine has emotions.

4.3.2. Participants' Attitudes toward Emotional Messages

The participants had a mean score of ≥3.0 for items regarding their interest toward emotional messages and the effects of emotional messages. Moreover, the modal score for these items was satisfactory. However, the participants generally did not pay sufficient attention to emotional messages. This result implies that the influence of messages on users is determined by how interested users are in these messages. A high score on the effects of emotional messages indicates the system's effectiveness in arousing user interest. Therefore, the effectiveness of emotional messages can be enhanced by increasing user interest in them.

4.3.3. Effects of Crucial Notifications on User Behavior

The view rate of crucial notifications was 76.6%, and instruction execution rate was 68.1%. These results indicated that user behavior was affected by the emotions of the system. The instruction execution rate was high because most of the participants were convinced that their machine had emotions, and they were willing to reflect on the emotional messages they received.

4.3.4. Research Limitations

This study recruited students enrolled in the general education courses of a university as the research participants. Therefore, the study results could only reflect the characteristics of the student population in the university and the region where the university is located. To address this limitation, future scholars should increase their sample size and diversify the participants in their research. Additionally, the system used in this study is only operable using Android devices and does not work on iOS devices. Therefore, future designs should account for compatibility with iOS devices. This will allow more users to operate the proposed system.

Picard posited that for machines to possess emotions, they must contain the following components: emergent emotion and emotional behavior, fast primary emotion, cognitive-generated emotion, emotional experiences, cognitive awareness, physiological awareness, subjective feelings, and body–mind interaction. Accordingly, this study employed these components to construct the proposed system. Picard also asserted standards should be developed to evaluate the performance of these components. The present researchers will further explore such standards in their future studies.

5. Conclusions

The experimental findings of this study indicate that compared with the participants' interest in emotional messages, the degree of attention paid by them to emotional messages more substantially affected their perception of whether their machine had emotions. Therefore, developers should focus on enhancing such attention to make users more willing to receive emotional messages or even click on the agent voluntarily to receive emotional messages. After attracting user attention, program designers should enhance the influence of emotional messages on users by appropriately designing the content of these messages, thereby persuading users that the emotion system gives their machine the ability to express emotions. The frequency of push notifications should not be excessively high. If this frequency is excessively high, users' willingness to receive emotional messages may decrease. In the future, researchers should first design an effective method for emotional expression and then appropriately design the content of emotional messages. Moreover, the system should affect user behavior by convincing users that their machine has emotions. Finally, the degree of attention paid to emotional messages determines the quality of an emotion system, and researchers and designers should bear this in mind. In future, we hope that this system can be used in medicine to help people with long-term emotional distress achieve "micro-intervention" psychotherapy to improve their mood. "Micro-interventions," such as breathing training and visualization, allow the subjects to use various adopted or more modern psychotherapy practice modes [36]. Further, it helps the elderly achieve a better experience in using smartphones and a sense of intimacy similar to younger people.

Author Contributions: Formal analysis, H.-C.K.L.; Data curation, Y.-C.M.; Writing—original draft, H.-C.K.L., Y.-C.M. and M.L.; Writing—review & editing, H.-C.K.L., Y.-C.M. and M.L.; Methodology, Y.-C.M. All authors have read and agreed to the published version of the manuscript.

Funding: This research received no external funding

Conflicts of Interest: The authors declare there are no conflict of interest.

Appendix A

Table A1. System usability scale.

Items	Descriptions	-				
1	I think that I would like to use this system frequently.	1	2	3	4	5
2	I found the system to be unnecessarily complex.	1	2	3	4	5
3	I thought the system was easy to use.	1	2	3	4	5
4	I think that I would need the support of a technician to use this system.	1	2	3	4	5
5	I found the various functions in this system to be well integrated.	1	2	3	4	5
6	I think that excessive inconsistency exists in this system.	1	2	3	4	5
7	I believe that most people would learn to use this system very quickly.	1	2	3	4	5
8	I found the system to be very cumbersome to use.	1	2	3	4	5
9	I felt very confident using the system.	1	2	3	4	5
10	I had to learn many things before I could begin using this system.	1	2	3	4	5

Table A2. Research questionnaire.

Items	Descriptions	-				
1	When the smartphone expresses emotions, I feel as if the smartphone is alive.	1	2	3	4	5
2	Irrespective of whether the smartphone has emotions, I have the same view of the smartphone.	1	2	3	4	5
3	I care about the emotions of the smartphone.	1	2	3	4	5
4	Every time I turn on the smartphone, I often observe the emotions of the smartphone.	1	2	3	4	5
5	Even if I see the system's emotion icon, I may not always check its emotions.	1	2	3	4	5
6	After closing other apps, if I see the agent of the system, I pay attention to the emotions of the smartphone.	1	2	3	4	5
7	I turn on the smartphone specifically to check emotional messages.	1	2	3	4	5
8	When I think about the emotions of the smartphone, I do not turn on the smartphone.	1	2	3	4	5
9	Sometimes, I turn on the smartphone simply to check the agent's emotions.	1	2	3	4	5
10	I pay attention to the messages and emotional statements of the system.	1	2	3	4	5
11	I am not interested in the messages and emotional statements of the system.	1	2	3	4	5
12	I am curious about the emotional state of the system.	1	2	3	4	5
13	After reading the message of the system, I think about what makes the smartphone produce such emotions.	1	2	3	4	5
14	Even if I read the messages of the system, I do not understand the reason why the mobile smartphone produces such emotions.	1	2	3	4	5
15	After reading the information of the system, I can understand why the smartphone has negative or positive emotions.	1	2	3	4	5
16	The information provided by the system indicates the ways in which I should avoid using my smartphone.	1	2	3	4	5
17	The information provided by the system does not cause me to change the way I use my smartphone.	1	2	3	4	5
18	I adjust how I use my smartphone according to the information provided by the system and guess the emotions that the smartphone may generate.	1	2	3	4	5

References

1. Tang, G.; Shi, Q.; Zhang, Z.; He, T.; Sun, Z.; Lee, C. Hybridized wearable patch as a multi-parameter and multi-functional human-machine interface. *Nano Energy* **2021**, *81*. [CrossRef]
2. Gadzama, W.; Joseph, B.; Aduwamai, N. Global smartphone ownership, internet usage and their impacts on humans. *J. Commun. Net.* **2017**, *1*. Available online: http://www.researchjournali.com/view.php?id=3876 (accessed on 26 January 2021).
3. Taylor, K.; Silver, L. *Smartphone Ownership is Growing Rapidly around the World, but Not Always Equally*; Pew Research Center: Washington, DC, USA, 2019.
4. An, T.; Anaya, D.V.; Gong, S.; Yap, L.W.; Lin, F.; Wang, R.; Yuce, M.R.; Cheng, W. Self-powered gold nanowire tattoo triboelectric sensors for soft wearable human-machine interface. *Nano Energy* **2020**, *77*. [CrossRef]
5. Xiao, G.; Ma, Y.; Liu, C.; Jiang, D. A machine emotion transfer model for intelligent human-machine interaction based on group division. *Mech. Syst. Signal Process.* **2020**, *142*. [CrossRef]
6. Liang, Z.; Oba, S.; Ishii, S. An unsupervised EEG decoding system for human emotion recognition. *Neural Net.* **2019**, *116*, 257–268. [CrossRef]
7. Kaklauskas, A.; Kuzminske, A.; Zavadskas, E.K.; Daniunas, A.; Kaklauskas, G.; Seniut, M.; Raistenskis, J.; Safonov, A.; Kliukas, R.; Juozapaitis, A.; et al. Affective tutoring system for built environment management. *Comput. Educ.* **2015**, *82*, 202–216. [CrossRef]
8. Dinakaran, K.; Ashokkrishna, E.M. Efficient regional multi feature similarity measure based emotion detection system in web portal using artificial neural network. *Microprocess. Microsyst.* **2020**, *77*. [CrossRef]
9. Górriz, J.M.; Ramírez, J.; Ortíz, A.; Martínez-Murcia, F.J.; Segovia, F.; Suckling, J.; Leming, M.; Zhang, Y.-D.; Álvarez-Sánchez, J.R.; Bologna, G.; et al. Artificial intelligence within the interplay between natural and artificial computation: Advances in data science, trends and applications. *Neurocomputing* **2020**, *410*, 237–270. [CrossRef]
10. Shi, Y.; Zhang, Z.; Huang, K.; Ma, W.; Tu, S. Human-computer interaction based on face feature localization. *J. Vis. Commun. Image Represent.* **2020**, *70*. [CrossRef]
11. Feng, C.M.; Park, A.; Pitt, L.; Kietzmann, J.; Northey, G. Artificial intelligence in marketing: A bibliographic perspective. *Australas. Mark. J.* **2020**. [CrossRef]
12. Picard, R.W. Affective computing: From laughter to IEEE. *IEEE Trans. Affect. Comput.* **2010**, *1*, 11–17. [CrossRef]
13. Yadegaridehkordi, E.; Noor, N.F.B.M.; Ayub, M.N.B.; Affal, H.B.; Hussin, N.B. Affective computing in education: A systematic review and future research. *Comput. Educ.* **2019**, *142*. [CrossRef]
14. Poria, S.; Cambria, E.; Bajpai, R.; Hussain, A. A review of affective computing: From unimodal analysis to multimodal fusion. *Inf. Fusion* **2017**, *37*, 98–125. [CrossRef]
15. Calvo, R.A.; D'Mello, S.; Gratch, J.; Kappas, A. *The Oxford Handbook of Affective Computing*; Oxford University Press: Oxford, UK, 2015.
16. Nalepa, G.J.; Kutt, K.; Bobek, S. Mobile platform for affective context-aware systems. *Future Gener. Comput. Syst.* **2019**, *92*, 490–503. [CrossRef]
17. Shi, Q.; Zhang, Z.; Chen, T.; Lee, C. Minimalist and multi-functional human machine interface (HMI) using a flexible wearable triboelectric patch. *Nano Energy* **2019**, *62*, 355–366. [CrossRef]
18. Mao, X.; Li, Z. Agent based affective tutoring systems: A pilot study. *Comput. Educ.* **2010**, *55*, 202–208. [CrossRef]
19. Sundström, P.; Ståhl, A.; Höök, K. A user-centered approach to affective interaction. *Lect. Notes Comput. Sci.* **2005**, *1*, 931–938.
20. Ståhl, A.; Höök, K.; Svensson, M.; Taylor, A.S.; Combetto, M. Experiencing the affective diary. *Pers. Ubiquitous Comput.* **2009**, *13*, 365–378. [CrossRef]
21. Tsutsumi, D.; Gyulai, D.; Takács, E.; Bergmann, J.; Nonaka, Y.; Fujita, K. Personalized work instruction system for revitalizing human-machine interaction. *Procedia CIRP* **2020**, *93*, 1145–1150. [CrossRef]
22. Bretan, M.; Hoffman, G.; Weinberg, G. Emotionally expressive dynamic physical behaviors in robots. *Int. J. Human Comput. Stud.* **2015**, *78*, 1–16. [CrossRef]
23. Bates, J. The role of emotion in believable agents. *Commun. ACM* **1994**, *37*, 122–125. [CrossRef]
24. Ushida, H.; Hirayama, Y.; Nakajima, H. Emotion model for life-like agent and its evaluation. In Proceedings of the 15th National Conference on Artificial Intelligence and 10th Innovative Applications of Artificial Intelligence Conference, Madison, WI, USA, 26–30 July 1998; pp. 62–69.
25. Maria, K.A.; Zitar, R.A. Emotional agents: A modeling and an application. *Inf. Softw. Technol.* **2007**, *49*, 695–716. [CrossRef]
26. Posner, J.; Russell, J.A.; Peterson, B.S. The circumplex model of affect: An integrative approach to affective neuroscience, cognitive development, and psychopathology. *Dev. Psychopathol.* **2005**, *17*, 715–734. [CrossRef]
27. Savery, R.; Weinberg, G. A survey of robotics and emotion: Classifications and models of emotional interaction. In Proceedings of the 2020 29th IEEE International Conference on Robot and Human Interactive Communication (RO-MAN), Naples, Italy, 31 August–4 September 2020; pp. 986–993.
28. Balzer, W.; Moulines, C.U.; Sneed, J.D. *Structuralist Knowledge Representation: Paradigmatic Examples*; Rodopi: Amsterdam, The Netherlands, 2000.
29. Woodworth, R.S.; Schlosberg, H. (Eds.) *Experimental Psychology*; Holt, Rinehart and Winston: New York, NY, USA, 1954.
30. Russell, J.A. A circumplex model of affect. *J. Personal. Soc. Psychol.* **1980**, *39*, 1161. [CrossRef]
31. Bradley, M.M.; Greenwald, M.K.; Petry, M.C.; Lang, P.J. Remembering pictures: Pleasure and arousal in memory. *J. Exp. Psychol. Learn. Mem. Cogn.* **1992**, *18*, 379. [CrossRef]

32. Watson, D.; Tellegen, A.; Weinberg, G. Toward a consensual structure of mood. *Psychol. Bull.* **1985**, *78*, 1–16. [CrossRef]
33. Brooke, J. SUS: A quick and dirty usability scale. In *Usability Evaluation in Industry*; CRC Press: Boca Raton, FL, USA, 1995; Volume 189.
34. Alepis, E.; Virvou, M. Automatic generation of emotions in tutoring agents for affective e-learning in medical education. *Expert Syst. Appl.* **2011**, *55*, 9840–9847. [CrossRef]
35. Bohn, N.; Kundisch, D. What are we talking about when we talk about technology pivots?—A Delphi study. *Inf. Manag.* **2020**, *57*. [CrossRef]
36. Meinlschmidt, G.; Tegethoff, M.; Belardi, A.; Stalujanis, E.; Oh, M.; Jung, E.K.; Kim, H.C.; Yoo, S.S.; Lee, J.H. Personalized prediction of smartphone-based psychotherapeutic micro-intervention success using machine learning. *J. Affect. Disord.* **2020**, *264*, 430–437. [CrossRef]

Gaining a Sense of Touch. Object Stiffness Estimation Using a Soft Gripper and Neural Networks

Michal Bednarek *, Piotr Kicki, Jakub Bednarek and Krzysztof Walas

Institute of Robotics and Machine Intelligence, Poznan University of Technology, 60-965 Poznan, Poland; piotr.kicki@put.poznan.pl (P.K.); jakub.bednarek@put.poznan.pl (J.B.); krzysztof.walas@put.poznan.pl (K.W.)
* Correspondence: michal.bednarek@put.poznan.pl

Abstract: Soft grippers are gaining significant attention in the manipulation of elastic objects, where it is required to handle soft and unstructured objects, which are vulnerable to deformations. The crucial problem is to estimate the physical parameters of a squeezed object to adjust the manipulation procedure, which poses a significant challenge. The research on physical parameters estimation using deep learning algorithms on measurements from direct interaction with objects using robotic grippers is scarce. In our work, we proposed a trainable system which performs the regression of an object stiffness coefficient from the signals registered during the interaction of the gripper with the object. First, using the physics simulation environment, we performed extensive experiments to validate our approach. Afterwards, we prepared a system that works in a real-world scenario with real data. Our learned system can reliably estimate the stiffness of an object, using the Yale OpenHand soft gripper, based on readings from Inertial Measurement Units (IMUs) attached to the fingers of the gripper. Additionally, during the experiments, we prepared three datasets of IMU readings gathered while squeezing the objects—two created in the simulation environment and one composed of real data. The dataset is the contribution to the community providing the way for developing and validating new approaches in the growing field of soft manipulation.

Keywords: machine learning; tactile sensing; perception for grasping

1. Introduction

Humans have an innate ability to perceive the physics of the world around them. As we are biologically equipped with a very sophisticated sensory system that delivers data to the brain, no-one consciously plans how to grab a cup of tea, squeeze a wet sponge or flip a book page. We all know how to do that and how to predict deformations of different objects based on their physical properties. Moreover, humans have at their disposal soft and highly effective grippers—hands. Taking into account our assumptions about the world that come from our minds, combined with the embodied intelligence [1] of our hands, we can flawlessly adjust the process of manipulation to fluctuating external conditions. However, machines do not have such in-built proficiency. Thus, their ability to manipulate only allows for handling repetitive tasks and prevents them from adapting to new types of objects efficiently.

Biologically inspired soft grippers [2–5] are designed to handle not only rigid bodies but also deformable and usually delicate objects. How they interact with the real world and how they adjust to different objects is ruled by their property called *intelligence by mechanics* [1]. One can observe a significant rise in the number of available applications of sensors capable of capturing high-dimensional deformations of soft and unpredictable physical objects [6–8]. However, in our work, we state that traditional and widespread sensors based on microelectromechanical systems can also be successfully used to predict the physical nature of the robot's surroundings. Thereby, we propose a hybrid approach that connects an *embodied intelligence* of a soft gripper with an *artificial intelligence* system to provide an easy to use, open-source and inexpensive method of estimating the physical properties of objects with various stiffness parameters.

The following study presents the deep learning, real-world application for stiffness coefficient estimation based on data from Inertial Measurement Units (IMUs) attached to the fingers of the gripper. Our contributions are:

1. Creation of simulated environments for generating contact signals from IMU and examining the soft gripper in various scenarios.
2. Verification of the performance of three neural networks in the task of stiffness parameter estimation—one purely convolutional and two recurrent models.
3. The real-world verification of the proposed solution.
4. The extensive examination of the reality-gap between the simulated and real data.
5. The open-source implementation and data used in the experiments available online (https://github.com/mbed92/soft-grip).

To prepare the real-world experiment, we used a two-finger gripper based on the Yale OpenHand Project [2] with two IMUs attached to its fingers. The motivation behind the choice of that type of sensor is twofold. First of all, typically soft grippers have no hinges and do not use encoders; therefore, we cannot track their movement directly. Following the research on the Pisa/IIT SoftHand [9], the the IMU measurements are sufficient for the motion tracking of underactuated and elastic fingers of the gripper. Secondly, IMUs are inexpensive, small and widespread among the robotics community. We than replicated this setup in simulation to obtain more learning data and the control over generated signals.

The course of the research is shown in Figure 1 and consists of the following stages: first, for the set of deformable objects of different shape and different stiffness parameters we performed squeezing motion both in simulation and in the real world. In both cases we were registering IMU data. For both approaches we trained and tested three different neural networks architecture. The final outcome of the learning process were estimated/regressed stiffness parameters of the objects. We started our investigation with experiments carried out exclusively on data from the physics simulator to verify the capabilities of three different architectures and examine the generalisation of the stiffness parameter regression between different shapes of squeezed objects. Thereafter, we investigated the problem of closing the reality gap between the simulation and real-world data. In our experiments, we exploited the MuJoCo [10] simulator to provide a sufficient number of training samples. The IMU device model used in our work was the MPU-9250 model. In Figure 2, there is presented the setup used in the real-world scenario with its simulation model and exemplary objects.

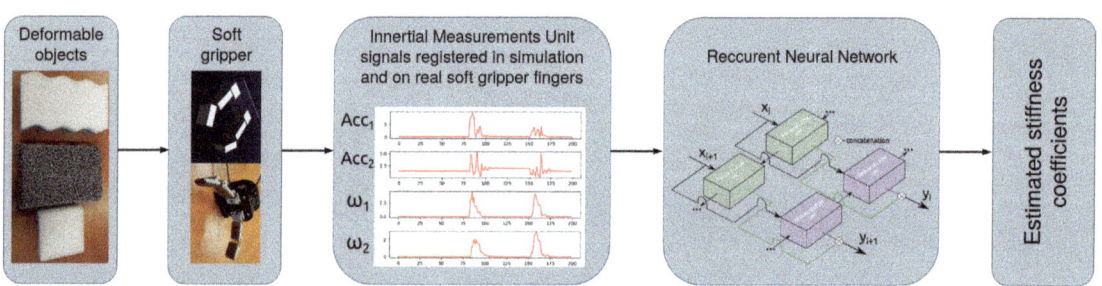

Figure 1. Schematic diagram of the proposed system. The set of deformable objects is squeezed in simulation and with the real gripper. The data is registered from Inertial Measurements Units mounted on the gripper fingers. Recurrent Neural Network is performing deformable object stiffness coefficient regression based on registered data from IMU.

The remainder of the paper organised as follows. First, we will review the state-of-the-art in the field of physical parameters estimation from haptic data. Then, we will provide a description of prepared setups and our experiments. Next, we will move on to the results section followed by the discussion. Finally, concluding remarks will be provided.

Figure 2. To test our system, we arranged a real-world scenario using a 2-finger Yale OpenHand gripper [2]. To provide a sufficient number of training samples for the learning process, we modelled the gripper in the MuJoCo simulator as it is depicted in (**a**). In (**b**), real fingers consist of three plastic blocks with flexible parts made of urethane. In (**c**), there are presented examples of sponges, exposing different stiffness, used in our real-world experiments.

2. Related Work

In this section, we provided a comprehensive literature review both on the approaches to measuring and estimating the object stiffness. Further, we showed current advances in processing data from IMUs for a wide range of purposes. Finally, we presented a brief overview of underactuated grippers with an emphasis on the soft grippers.

2.1. Measuring and Estimating a Stiffness

The knowledge about material's stiffness is highly demanded in many practical applications such as industrial robotics, where a robot may use this information to predict an object's deformation. We present current advances in finding object stiffness in two general approaches: *measurement*, where the result was obtained with the usage of advanced, dedicated sensors and *estimation*, where we focused on the possible use of all available information relevant for a given task.

Measurement—The practical application of the stiffness measurement was shown in [11], where the authors proposed a method for continuous rail rigidity measurements using the accelerometer and oscillating mass on the rolling wheel. This indicates that the issue under examination is of great importance in engineering. Unlike the previous method, the noncontact measurement of spindle stiffness was presented in [12]. The authors proposed a magnetic loading device that enables one to perform the measurement while the spindle rotates. Due to the usage of magnetic loading, that method is limited to the ferromagnetic objects. Measuring the stiffness is also possible at a much smaller scale. The authors of [13] presented the review of the nanoindentation continuous stiffness measurement technique and its applications. The range of stiffness coefficients of materials is extensive. To avoid saturation and enhance precision, authors of [14] proposed a portable measurement tool able to adjust the sensing range by manipulating tool parameters, such as touch module separation, indenter protrusion and spring constant of the force sensing module. Authors of [15,16] analysed the stiffness measurement techniques applied to the polymer foams, which are cognate to those used in this paper. In [15] a procedure for measuring the stiffness of the object using dot markers on the object and compression plates to exert the force on the object was proposed. Authors stress the fact that nonaxial compression tests result in worse performance, but it is usually the case in robotic manipulation. As in our method, in [9] authors proposed the IMU-based approach but in a different task—the reconstruction of the configuration of a soft gripper. As opposed to that work, we propose to indirectly measure the stiffness property by the change of behaviour of the soft gripper while squeezing, not the gripper's configuration itself.

Estimation—The method for the object stiffness estimation from the force sensor readings was proposed in [17]. Authors used small optical force sensors mounted on the fingertips, a known kinematic model of the robotic hand and a vision system to calculate the stiffness based on the force and displacement readings. An alternative approach that does not require measuring the object deformation was proposed in [18]. The authors proposed the Candidate Observer-Based Algorithm, which exploits two force observers, with different stiffness candidates, for estimating the stiffness of objects with complicated geometry. Unfortunately, the authors did not refer their method to the ground truth stiffness measurements. However, such a comparison was made in [19], where the neural network was trained to predict the stiffness coefficient based on the maximum penetration and the maximum contact pressure variation. An alternative deep learning approach for understanding the haptic properties of objects was proposed in [20]. The real-world objects were classified in the set of haptic adjectives in the multilabel fashion based on haptic signals from BioTac sensors [21] and images. That work shows that there exists a correlation between haptic sensor readings and the structure of the real-world objects, and in our work, we took advantage of that fact.

The extensive overview of machine learning methods in the soft robotics aspect is described in [22]. In the context of sensing, the authors distinguish sensor characterization and systems characterization. In the group of sensors characterization, the use of Recurrent Neural Networks for parameters regression is widespread, as we are dealing with signals and continuous values of sensor parameters. On the system characterization level, we are more focused on higher-level labels successful grasp [23] or slip detection. The use of the classification of signals with categorical values is more common. The more focused approaches are shown in [24], where learned control mechanisms were used, reinforcement learning [25] or learned differentiable models [26].

2.2. IMU Measurements Applications

The popularity of IMU usage stems from its widespread availability at a low price. One possible use in the robotics community is a robot's state estimation. In [27], acceleration and angular velocities collected from sensors located on the humanoid leg, together with joints positions were used to estimate the velocity of the robot links. Authors in [28] presented multiple interesting approaches to measure the ground reaction forces indirectly during the human walk with the use of wearable IMUs. The other field where the measurements of acceleration can be utilised is a material classification. In [29] authors used the haptic device SensAble Phantom Omni [30] to gather the accelerations and velocities while scratching the material surfaces. That dataset was used in [31], where a deep convolutional neural network was taught to map raw signals input to classes of textures. The presented method stays close to our solution. However, in our work, we performed regression instead of classification.

2.3. Underactuated and Soft Grippers

As the approach proposed in this paper requires a soft gripper, we present a brief overview of existing underactuated and soft grippers.

Underactuated grippers was an area of research for many years. One of the first grippers was designed by Tomovic and Boni in [32]. They proposed an anthropomorphic underactuated hand with five fingers and 14 joints, driven by a tendon-driven mechanism and two servo motors. Using tendons in the gripper designs provides an ability to easily adjust the gripper's shape to the object in a gentle way. To achieve grasps available only to the fully-actuated mechanisms, authors of [33] proposed an underactuated gripper with electrostatic brakes in joints, which enable to carry heavy objects by reducing power consumption and motor torque during a steady grasp. Nowadays, hybrid approaches, which combines fully actuated fingers for precise manipulation and underactuated ones for power grasp and compliance, are becoming popular [34,35]. A different hybrid approach is presented in [36], where authors proposed an underactuated gripper with a suction cup for picking up various objects in different working environments. Such hybrid solution allows for building multifunction robotic cells [37] for maximising the production rate.

On the other hand, for full compliance and shape adaptability, there is a lot of research about a special group of underactuated grippers—soft grippers [38]. A popular way of designing soft grippers is using elastomer actuators. In [39] authors used rubber fingers driven by the pressure in the chambers located inside the fingers. However, as used materials are usually soft, they are susceptible to damage. In response to that, authors of [40] presented usage of self-healing materials to construct a soft gripper able to repair itself. A different approach to control soft grippers is to use dielectric [41] or shape-memory based [42] actuators. However, probably the most popular group of soft grippers are those with passive structure driven by the external motors, such as adaptive compliant gripper proposed in [43] or biomimetic soft-hand [44]. To this group also belongs a Yale Hand gripper [2], which we used in our research. It is a low-cost two-fingered open-source underactuated robotic hand, which is built with 3D-printed components with compliant, flexible joints, and driven by tendons actuated with servos.

Interested readers may refer to recent more comprehensive underactuated and soft grippers reviews such as [34,38,45].

3. Method

In the following section, we described the experimental design and provided detailed information about both real-world and simulated environments for our experiments. Then, we described the proposed neural network architectures used in our research.

3.1. Real Data

The Yale OpenHand shown in Figure 2 is the underactuated, two-finger soft gripper with joints in the form of urethane elements to assure the elasticity of fingers. The real-world model was 3D printed and driven by hobby servos capable of generating a force up to 10 N. The IMUs were mounted at the fingertips of the hand. The IMU readings were used to estimate grasped objects stiffness. In our work, we assessed how the embodied intelligence of such soft gripper could be used alongside with the artificial intelligence system to predict the real stiffness coefficient of a squeezed object. In the following paragraph, the real-world data gathering process was presented.

First we estimated the stiffness coefficient for real objects in the dataset. To calculate ground-truth values of the stiffness coefficient of real-world objects, we used the Universal Robot UR3 collaborative manipulator, which was able to measure torques and forces in its joints and tool respectively. The robot had 3D printed plastic bar mounted at the flange. Using the Dynamic Force Control mode and pressing objects with the desired force, we were able to accurately measure the displacement under specific force from robot state readings. Thus, the stiffness parameter was computed according to Equation (1), where f_1 and f_2 are forces in Z-axis while pressing an object with a tool and $|d_1 - d_2|$ is the relative distance that correspond to the deformations under f_1 and f_2. We are aware that chosen objects express nonlinear behaviour in their stiffness model (e.g., the greater robot compress the sponge, the less deformation it adds). However, objects did not reflect that nonlinear effects in the specified range of exerted forces. Therefore, in our work, we assumed that the estimated stiffness parameter is homogeneous for the entire object. Table 1 contains stiffness coefficients measured experimentally for each object.

$$k = \frac{|f_1 - f_2|}{|d_1 - d_2|} \quad (1)$$

Table 1. Stiffness coefficients computed for 5 different real objects.

Object	Stiffness [N/m]
Wire sponge	909
Hard sponge	1020
Polish sponge	735
Soft sponge	380
Squash ball	1353

After measuring the value of the ground-truth stiffness coefficients, we used Yale OpenHand to perform squeezing motion of each object and collected IMU readings during motion execution. In total, we gathered 500 series. They consist of 12 sensor readings (2 · IMU readings: $[a_x, a_y, a_z, \omega_x, \omega_y, \omega_z]$) each 200 time steps long. All samples are equally distributed among the objects–100 samples per each object. The data was split into two subsets—200 train and 300 test samples that were used in sim-to-real experiments. Both sets in all our experiments remain unchanged. Thus, test data is never used in the NN training. To address the issue of a physical interpretation of obtained stiffness, taking as input the accelerations and angular velocities, we claim that the motion of gripper fingers registered while squeezing different objects would vary significantly, which was presented in Figure 3. One can observe that depending on the object's stiffness the magnitude and oscillations of both-angular velocity ω and linear accelerations Acc were significantly different from each other, e.g., in the range of values or the oscillation rate. Taking that phenomenon into account we put forward the thesis that it is possible to distinguish between different stiffness parameters in the space of IMU sensors registered during squeezing of these objects.

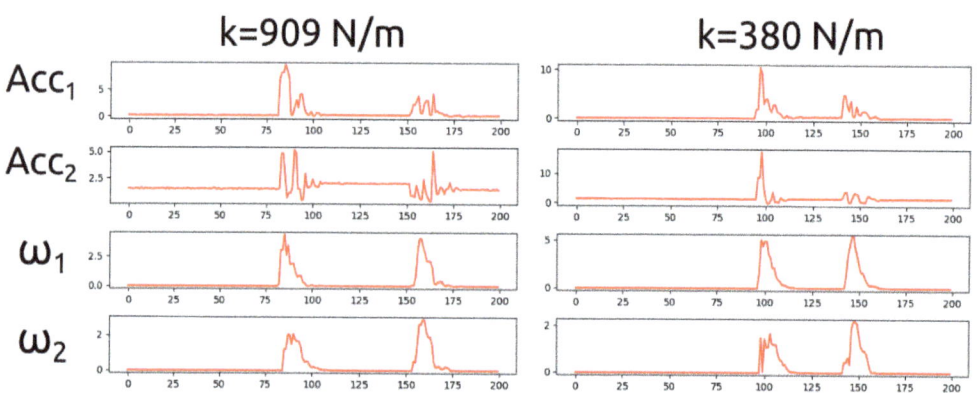

Figure 3. Comparison between exemplary samples from the real-world dataset while squeezing objects with different stiffness values with a soft gripper. Values a_1, a_2, ω_1 and ω_2 refers to the magnitude of registered accelerations and angular velocities and are expressed in $\frac{m}{s^2}$ and $\frac{rad}{s}$ respectively.

However, in our real-world dataset, there were only five different objects with distinct stiffness values, which served as labels. In that situation, the number of different labels was not sufficient to perform a successful regression. In fact with such a low diversity, a regression would inevitably turn into a classification and that was not desired in the task of stiffness estimation. To overcome that problem, we prepared a second dataset based on the simulation, where there was a possibility to generate more training samples. Stiffness coefficients were adjusted to meet measured values.

3.2. Simulation

Modern neural networks frequently suffer from the limited ability to generalise to new domains which are out of their training dataset. However, the rising popularity of machine learning techniques in the robotics community leads to a significantly increased need for data from a variety of experiments. To fulfil that demand, the state-of-the-art approach is to perform experiments in simulation and use them to feed neural networks. In the case of tasks which involve physical interactions, researchers can choose from a wide range of available physics simulators. In our case, we selected MuJoCo physics simulator, due to its new features regarding soft objects modelling. The simulated soft-robotic gripper was shown in Figure 4. Fingers were connected by tendons and they are pulled by the actuator, which simulates the pneumatic cylinder. Our model was based on the three-finger real gripper [3] but with one finger removed. As it is depicted in Figure 4a, during experiments, our gripper squeezed and released objects of three shapes—a ball, box and a cylinder, all with a variable stiffness parameter. To simulate elastic deformations of the gripper, each geometrical block of each finger is connected to others by three hinges. In this setup, we can easily adjust the ranges of each joint in a roll, pitch and yaw axes, as was depicted in Figure 4b. Finally, each 8-block finger behaves similarly to the elastic finger.

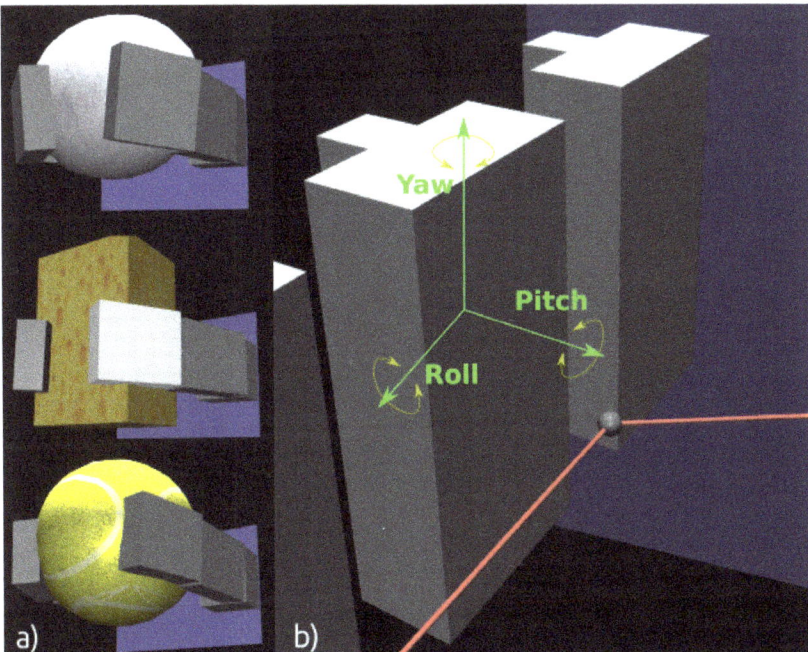

Figure 4. Soft-robotic gripper in the MuJoCo environment: (**a**) the gripper squeezes and releases objects in three shapes—a ball, a box and a cylinder, all with a variable stiffness parameter; (**b**) each geometrical block of each finger is connected to others by three hinges. In this setup, we can easily adjust the ranges of each joint in a roll, pitch and yaw axes.

A stiffness coefficient in all our experiments (including the real-world scenario) is defined in the same way as in the MuJoCo simulator—the *softness* of an object is characterised as the stiffness of springs attached from one side to the geometrical blocks on a surface and from the other side to the centre of it. We always assume that the object is homogeneous.

The data collection was performed using the following steps. An object is located between fingers and the actuator starts to close the gripper to squeeze the object. After a half of an episode, the gripper opens. During the process, an object is embraced by fingers that adapt themselves to its shape. A stiffness coefficient is expressed in N/m and varies among episodes to equally cover the range (from 300 to 1400$\frac{N}{m}$), which fits the real-world data range. A mass of all parts was adapted to the real values, as well as the mechanical impedance of objects, damping, and stiffness of all joints and springs in the system. Two IMUs are mounted on a MuJoCo's element called *site* and located in the 3/4 of the length of each finger in the outside part of it. For experiments, we prepared two simulation datasets. The first one resembles the real-world data and consists of 5000 training-validation samples gathered from squeezing the box object only. We use it for an enrichment of real-world data. The second one was composed of objects in three different shapes—boxes, cylinders and spheres. It counts 3999 training-validation samples—1333 samples per each object. In our research, it was used to verify whether the NN can avoid overfitting to any particular shape. Additionally, to verify the NN performance among different shapes of objects we prepared three test datasets—133 samples for each object.

3.3. Experimental Design

The performance of Neural Networks (NN) was verified using a k-fold cross-validation technique in each experiment. That method assesses the error rate and the generalisation ability of predictive models. In our research, data is processed as follows: we shuffle the dataset, then split the dataset into k subsets (folds), proceed with training using the $k-1$ folds of data and validate the performance at the end of an epoch using the k-th fold. Additionally, unless otherwise stated, after each epoch, we test the current NN model using separate test data. After that, the procedure is repeated by starting the training of a neural network from scratch on other folds of data. In our research, to ensure a fair comparison of trained NNs, we did the 5-fold cross-validation for all experiments. As we perform the regression task, we chose a Mean Absolute Error (MAE) and a Mean Absolute Percentage Error (MAPE) as the performance metrics, to verify both absolute and relative errors. Considering the usage of the cross-validation technique, in the following description of datasets, we provided the number of samples in the training-validation sets together and separately for test sets if needed. The summary of all datasets used in our experiments was presented in Table 2.

Table 2. The number of samples in datasets used in our experiments based on the cross-validation.

Name	Train/Validation	Test
Simulation (box only)	5000	-
Simulation (all shapes)	3999	399
Real-world	200	300

3.4. Network Architecture

Our neural networks were predicting the stiffness parameter from fixed length sequences of accelerations and angular velocities measured by IMUs. In our research we proposed to test three types of neural networks—the ConvNet based entirely on 1D convolutional blocks, the ConvLSTMNet with forward LSTM units and the ConvBiLSTMNet with bidirectional LSTM units. In both cases of LSTM-based NNs models, the recurrent part is placed after the convolutional block. At the end of each architecture, we placed a fully-connected layer named the Regression Block. The scheme of proposed neural network architectures are depicted in Figure 5.

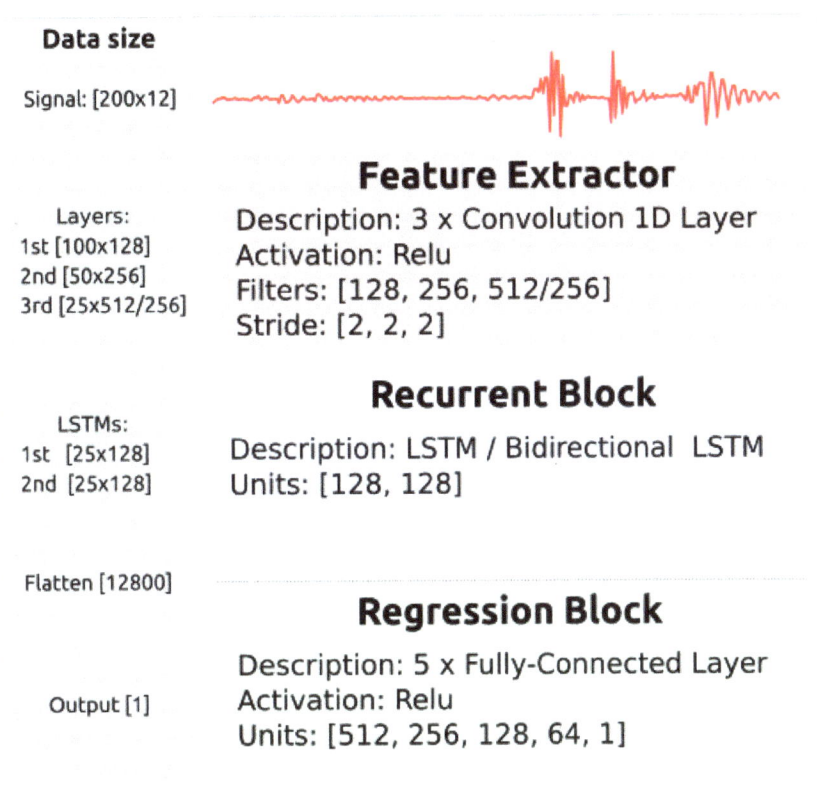

Figure 5. In our networks, the Feature Extractor produced high-level features from an input using 1D convolutions. In the ConvBiLSTMNet and the ConvLSTMNet, the Recurrent Block processed these features to find relevant connections for the stiffness estimation. However, in the former architecture it was done in the forward and backward manner (from the beginning of the signal and back). Finally, the Regression Block transformed high-level features into one scalar value. In our experiments, we exploited three architectures of neural networks. The difference is in the Recurrent Block—both recurrent NNs have the reduced number of filters in the last convolutional layer and added LSTM cells with 256 units (2 × 128), while in the ConvNet the output from the Feature Extractor is passed directly to the Regression Block.

Feature Extractor—The neural network input was a standardised sensor reading in the form of the two-dimensional tensor. Each sample consisted of 12 time series with a length of 200. The main task of that block is to extract features while remaining in the time domain. Hence, data could be further processed recurrently or passed to the Regression Block directly. The Feature Extractor consisted of three consecutive 1D convolution layers with strides equal 2. In the ConvNet the number of filters was set to 128, 256, 512, while in the ConvLSTMNet/ConvBiLSTMNet, the last convolution block was reduced to 256 filters and replaced by the recurrent block with the same size.

Recurrent Block—It processes high dimensional time series from the Feature Extractor in a recurrent manner using LSTM [46] or bidirectional LSTM cells [47]. The input is mapped to a fixed-length vector that represents the entire sensor reading in itself. In that way, we obtained a global, reduced description of the signal. Each recurrent cell consists of 128 units, as depicted in Figure 6. In the the ConvLSTMNet, both LSTM cells are organised in two sequential layers processing the input in the forward direction only. Outputs of that block are finally forwarded to the Regression Block.

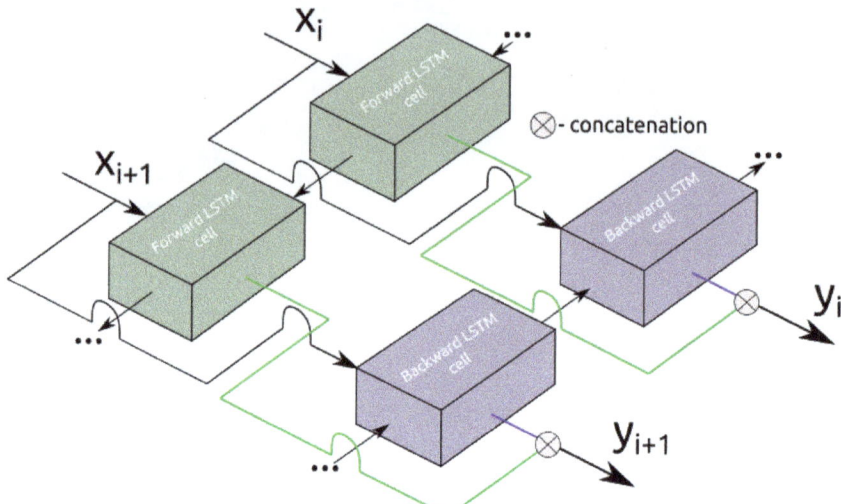

Figure 6. The core idea standing behind the bidirectional LSTM used in the ConvBiLSTMNet is as follows–to prevent losing a context by the cell, process a sequence from the beginning to the end, do the same in the reversed direction and concatenate both *passes*. Input x_i refers to the i-th feature vector returned by the convolutional block.

Regression Block—The last block was used to do a regression and output an estimated stiffness coefficient. The necessity of using a fully-connected block stems from the fact that extracted features and time dependencies between them are critical ingredients in the regression process, but they are not the answer itself. At the very end of the processing, it is necessary to transform the obtained features into stiffness coefficient estimate, which can be easily performed using the stack of fully-connected layers. The number of units in each layer remains unchanged for all tested architectures and is 512, 256, 128, 64, 1.

4. Results

The section of the results was divided as follows. Firstly, the simulation results were presented. We verified which NN yielded the best performance on the simulation datasets and how well it was able to generalise among different shapes of squeezed objects. After that, the real-world experiments were conducted using the NN architecture chosen during the simulation test stage. Finally, we focused on the closing of the reality gap between simulation and real-world data. In all experiments we validated or models using the k-fold cross-validation technique and provided results obtained for the best epoch per each fold according to the MAPE. To ensure a fair comparison, in all our experiments and tests we used Adam optimiser with a learning rate set to 0.001. Each model was trained with the batch size 100 and all our solutions were trained for 100 epochs per each fold of the cross-validation.

4.1. Neural Network Architecture Comparison

First of all, in our experiments, we compared three types of neural networks using data from simulation and chose the best one for further experiments. The values of MAE/MAPE metrics from the cross-validation procedure were presented in Table 3. The best performing network—the ConvBiLSTMNet, was chosen for further experiments.

Table 3. The comparison of three NN architectures according to MAE/MAPE metrics. The usage of bidirectional LSTM units gave an improved performance comparing to the ConvNet and the ConvLSTMNet.

	ConvNet		ConvLSTMNet		ConvBiLSTMNet	
k-Fold	MAE	MAPE	MAE	MAPE	MAE	MAPE
I	19.1	2.4	6.2	0.8	6.2	0.8
II	11.8	1.6	5.4	0.7	5.4	0.7
III	15.1	2.2	7.8	1.1	7.8	1.1
IV	14.6	1.9	6.7	0.9	6.7	0.9
V	18.1	2.1	6.2	1.0	6.2	1.0
MEAN	15.7	2.0	6.8	0.9	6.5	0.9
SD	2.9	0.3	0.9	0.2	0.7	0.1

4.2. Shape Generalisation

To verify the capability of the ConvBiLSTMNet to successfully estimate the stiffness parameter we conducted more experiments using the simulation-only datasets. We started the cross-validation procedure from scratch for chosen model and reported the MAE/MAPE for three different datasets in Table 4. Each test dataset was composed of sensor readings from squeezing only one type of object so that the findings of the shape-dependent stiffness parameter regression could be provided.

Table 4. The results from experiments on shape-invariant estimation of the stiffness parameter using ConvBiLSTMNet.

	Dataset					
k-Fold	Ball		Box		Cylinder	
	MAE	MAPE	MAE	MAPE	MAE	MAPE
I	20.3	2.0	24.1	1.8	15.6	1.8
II	29.6	2.6	12.9	1.6	15.8	1.9
III	27.1	2.0	22.8	1.8	16.0	1.9
IV	21.8	2.1	17.7	16.6	18.4	1.9
V	19.3	2.0	24.4	1.5	20.8	1.9
MEAN	23.6	2.1	20.4	4.7	17.3	1.9
SD	4.5	0.3	5.0	6.7	2.2	0.0

4.3. Sim-To-Real Gap

The central part of our research was about assessing the reality gap in the task of the stiffness parameter estimation. In that part of the experiments, we performed 5 training procedures of the ConvBiLSTMNet on data with different number of real-life examples or noise added to simulation data, each composed of 5-fold cross-validation. In Table 5 we reported MAE/MAPE metrics gathered while *testing* each model on the separate dataset, not involved in the training/validation procedure. In the *sim + noise* experiment we tried to close the reality gap, by adding a zero-mean Gaussian noise with standard deviation set to $0.7 \frac{m}{s^2}$ for accelerations and $0.06 \frac{rad}{s}$ for the gyroscope readings. The parameters of the noise were adjusted by trials and errors, thus too large standard deviation resulted in the lack of the convergence ability of the NN, while too small caused model to overfit to the simulation data and no clear rule for that phenomena is known. Each next cross-validation turn was performed on simulation datasets without noise and with a small number N of real-world data samples included in the training part. In Table 5 we refer to them as *sim + N real*.

Table 5. MAE/MAPE results reported for best epochs from each of the cross-validation turns. Introducing to the network even a small number of real-world sensor readings resulted in a significant improvement in the performance.

Experiment Name	k-Fold										MEAN	
	I		II		III		IV		V		MAE	MAPE
	MAE	MAPE	MAE	MAPE	MAE	MAPE	MAE	MAPE	MAE	MAPE		
sim + noise	281.3	37.7	275.0	38.5	275.6	38.4	282.7	37.6	256.6	37.9	274.2 ± 10.4	38.0 ± 0.4
sim + 50 real	190.6	23.1	216.1	27.1	187.8	26.4	151.8	21.6	200.7	27.7	189.4 ± 23.8	25.2 ± 2.7
sim + 100 real	134.6	20.6	108.3	17.6	134.9	19.6	126.8	18.6	126.6	18.3	126.2 ± 10.8	18.9 ± 1.2
sim + 150 real	89.3	12.9	85.9	13.7	92.7	13.2	73.9	11.0	79.9	10.2	84.3 ± 7.5	12.2 ± 1.5
sim + 200 real	66.9	9.1	49.3	7.0	82.6	10.9	67.4	8.4	56.6	8.0	64.6 ± 12.6	8.7 ± 1.5

5. Discussion

In the following section, we summarised obtained results and our observations for three types of experiments carried out in the course of our research.

Architecture Choice—We compared the performance of three types of neural networks in the task of a stiffness parameter estimation from IMUs readings, to choose the best one for the further analysis. All models were examined on the simulation dataset without real-world data samples. In Table 3 one can observe the results from cross-validation on the simulation dataset. The mean results of the MAE/MAPE show the advantage of the LSTM-based models in the performed task. The conclusions are twofold. Firstly, the ConvBiLSTMNet is more accurate in its predictions than ConvNet, resulting in MAE of $6.5 \frac{N}{m}$ and MAPE of 0.9%, which means the improvement over $9.5 \frac{N}{m}$ and 1.1% achieved by the ConvNet. Secondly, the stability of the learning process also improved and it can be observed in deviations of errors obtained between cross-validation folds. For ConvNet the standard deviation of results is $2.9 \frac{N}{m}$ MAE and 0.3% MAPE, while the ConvBiLSTMNet decreased these values to $0.9 \frac{N}{m}$ and 0.2% respectively. Comparing two recurrent NNs, one can observe that the results are similar. However, the ConvBiLSTMNet exhibits better performance in the MAE, what means than on average it made a lesser absolute error, hence that architecture was chosen for further experiments.

Shape-Invariant Predictions—To verify the generalisation capability of the ConvBiLSTMNet and verify its performance on different types of objects, we performed additional experiments. In Table 4 we gathered the MAE/MAPE from testing the network on three separate datasets, each of which included only one shape of object, while training on all shapes at once. All the results suggest that the proposed NN was able to generalise among different types of shapes and perform the shape-invariant stiffness parameter prediction. It appears that the cylinder-shaped objects are the easiest in the performed task, which is reflected in the lowest errors $17.3 \frac{N}{m}$ MAE and 1.9% MAPE. However, box objects gave the smaller values of MAE ($20.4 \frac{N}{m}$) than ball-shaped objects ($23.6 \frac{N}{m}$), while looking at the MAPE the situation was the opposite—larger error was observed for boxes (4.7%/2.1%). This means that the NN was inaccurate more often while estimating large stiffness values for boxes that resulted in the increased relative metric (MAPE), while for ball-shaped objects the quality of the estimation was decreased for small values that gave increased absolute measure (MAE).

Closing The Reality Gap—In the task of haptic recognition of physical parameters, data from the physics simulator appeared to resemble the real-world IMU readings only to some restricted extent. Although the results from *sim + noise* tests were significantly worse than any of the *sim + real* trail, the mean MAPE 38% suggests that the correspondence between the simulation-only and real-world signals exists. Additionally, it is important to note that MAE/MAPE values from each fold in the *sim + noise* experiment remained relatively close to each other, which means that the model prediction performance was similar for the entire dataset, as it was equally balanced in the stiffness parameters range. However, the reality gap cannot be considered as a solved problem, because the greatest improvement was observed for experiments with the real-world sensor readings included in the training dataset. In Figure 7, one can observe the decreasing value of MAE/MAPE metrics as the number of real data samples are added to the training dataset. In our experiments we do not include the results from the training on the real-world data only, as they would be incomparable with other experiments, due to the low variability of the stiffness coefficient. Additionally, the number of data samples would be too low to assess the fair comparison in the real-world scenario. The lowest MAE/MAPE obtained in experiments on closing the reality gap were achieved for *sim + 200 real* trial and were equal to $64.6 \frac{N}{m}$ and 8.7%. However, in the *sim + 50 real* experiment, the added number of real samples constituted only 1.2% of the entire training dataset, but the largest performance improvement among all experiments was observed. The improvement was $84.8 \frac{N}{m}$ and 12.8% of the MAE/MAPE.

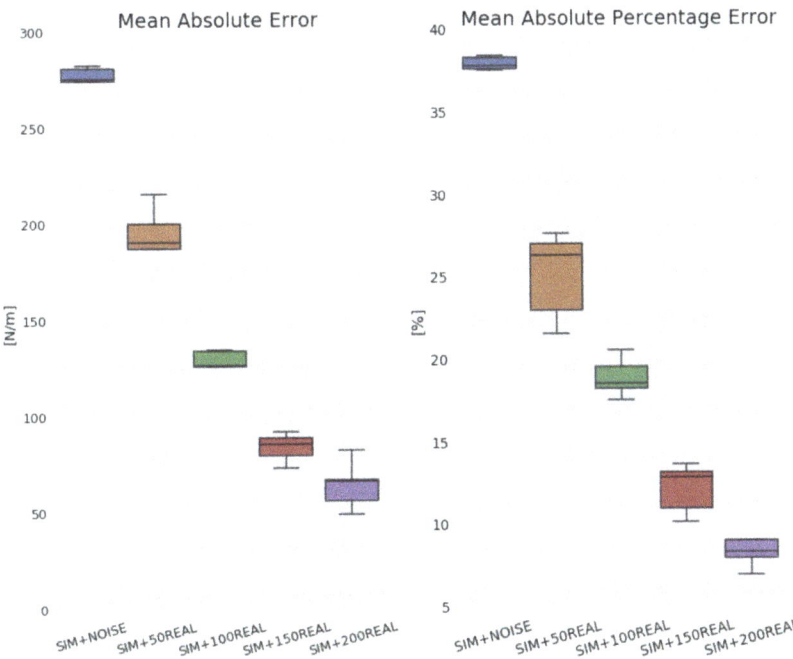

Figure 7. Results of MAE/MAPE from the testing on real-world data presented in the box plot. As the number of real data samples included in the training dataset increases, the test error decreases. Boxes represent consecutive experiments and consist of the five-number summary of the result (from the bottom of each box): minimum, first quartile, median, third quartile and maximum value.

6. Conclusions

We have shown that estimation of the object's physical parameters using data from IMU sensors is possible and beneficial due to the low cost of setup and no further need for sophisticated equipment. Our deep learning solution solves a problem of the stiffness estimation in the soft robotics area, introducing a novel approach, which associates an embodied and artificial intelligence. Their combination may lead to a system robust to unforeseen and changing external conditions. While currently used methods of stiffness search exploit techniques of measurement or direct estimation, the method proposed by us is characterised by the discovery of knowledge and causal relationships related to the characteristics of a given object and its physical features. Research on the discovery of knowledge acquired by neural networks may result in the diagnosis of the intuition behind the natural behaviour of humans in the tasks of manipulating objects. We find it likely that similar solutions, based on low-cost sensors and deep learning, may be successfully applied for robotic manipulation in everyday scenarios. We hope that the published data and the implementation of neural networks used in our experiments will inspire other researchers to delve into the research area of soft grippers and perception of the physical world based on tactile data in robotics.

Author Contributions: Conceptualization, M.B. and J.B.; Formal analysis, P.K.; Funding acquisition, K.W.; Methodology, M.B.; Software, M.B.; Supervision, P.K. and K.W.; Writing—original draft, M.B., P.K., J.B. and K.W.; Writing—review and editing, M.B. and K.W. All authors have read and agreed to the published version of the manuscript.

Funding: This work is supported by grant No. LIDER/3/0183/L-7/15/NCBR/2016 funded by The National Centre for Research and Development (Poland).

Conflicts of Interest: The authors declare no conflict of interest.

References

1. Pfeifer, R.; Lungarella, M.; Iida, F. Self-Organization, Embodiment, and Biologically Inspired Robotics. *Science* **2007**, *318*, 1088–1093. [CrossRef]
2. Odhner, L.U.; Ma, R.R.; Dollar, A.M. Open-Loop Precision Grasping with Underactuated Hands Inspired by a Human Manipulation Strategy. *IEEE Trans. Autom. Sci. Eng.* **2013**, *10*, 625–633. [CrossRef]
3. Homberg, B.S.; Katzschmann, R.K.; Dogar, M.R.; Rus, D. Haptic identification of objects using a modular soft robotic gripper. In Proceedings of the 2015 IEEE/RSJ International Conference on Intelligent Robots and Systems (IROS), Hamburg, Germany, 28 September–3 October 2015; pp. 1698–1705.
4. Li, S.; Stampfli, J.J.; Xu, H.J.; Malkin, E.; Diaz, E.V.; Rus, D.; Wood, R.J. A Vacuum-driven Origami "Magic-ball" Soft Gripper. In Proceedings of the International Conference on Robotics and Automation, ICRA 2019, Montreal, QC, Canada, 20–24 May 2019; pp. 7401–7408.
5. Manti, M.; Hassan, T.; Passetti, G.; D'Elia, N.; Laschi, C.; Cianchetti, M. A Bioinspired Soft Robotic Gripper for Adaptable and Effective Grasping. *Soft Robot.* **2015**, *2*, 107–116. [CrossRef]
6. Atalay, A.; Sanchez, V.; Atalay, O.; Vogt, D.; Haufe, F.; Wood, R.J.; Walsh, C.J. Batch Fabrication of Customizable Silicone-Textile Composite Capacitive Strain Sensors for Human Motion Tracking. *Adv. Mater. Technol.* **2017**. [CrossRef]
7. Chorley, C.; Melhuish, C.; Pipe, T.; Rossiter, J. Development of a Tactile Sensor Based on Biologically Inspired Edge Encoding. In Proceedings of the Advanced Robotics, ICAR, Munich, Germany, 22–26 June 2009.
8. Sie, A.; Realmuto, J.; Rombokas, E. A Lower Limb Prosthesis Haptic Feedback System for Stair Descent. In Proceedings of the 2017 Design of Medical Devices Conference, Minneapolis, MN, USA, 10–13 April 2017; V001T05A004.
9. Santaera, G.; Luberto, E.; Serio, A.; Gabiccini, M.; Bicchi, A. Low-cost, fast and accurate reconstruction of robotic and human postures via IMU measurements. In Proceedings of the 2015 IEEE International Conference on Robotics and Automation (ICRA), Seattle, WA, USA, 26–30 May 2015; pp. 2728–2735.
10. Todorov, E.; Erez, T.; Tassa, Y. MuJoCo: A Physics Engine for Model-Based Control. In Proceedings of the 2012 IEEE/RSJ International Conference on Intelligent Robots and Systems, Algarve, Portugal, 7–12 October 2012; pp. 5026–5033.
11. Wang, P.; Wang, L.; Chen, R.; Xu, J.; Xu, J.; Gao, M. Overview and Outlook on Railway Track Stiffness Measurement. *J. Mod. Transp.* **2016**, *24*, 89–102. [CrossRef]
12. Matsubara, A.; Yamazaki, T.; Ikenaga, S. Non-contact Measurement of Spindle Stiffness by Using Magnetic Loading Device. *Int. J. Mach. Tools Manuf.* **2013**, *71*, 20–25. [CrossRef]
13. Li, X.; Bhushan, B. A Review of Nanoindentation Continuous Stiffness Measurement Technique and Its Applications. *Mater. Charact.* **2002**, *48*, 11–36. [CrossRef]
14. Sul, O.; Choi, E.; Lee, S.B. A Portable Stiffness Measurement System. *Sensors* **2017**, *17*, 2686. [CrossRef]
15. Marter, A.; Dickinson, A.; Pierron, F.; Browne, M. A Practical Procedure for Measuring the Stiffness of Foam like Materials. *Exp. Tech.* **2018**, *42*, 439–452. [CrossRef]
16. Petrů, M.; Novák, O. Measurement and Numerical Modeling of Mechanical Properties of Polyurethane Foams. In *Aspects of Polyurethanes*; Yilmaz, F., Ed.; IntechOpen: Rijeka, Croatia, 2017; Chapter 4.
17. Kicki, P.; Bednarek, M.; Walas, K. Robotic Manipulation of Elongated and Elastic Objects. In Proceedings of the 2019 Signal Processing: Algorithms, Architectures, Arrangements, and Applications (SPA), Poznan, Poland, 23–25 September 2019; pp. 23–27.
18. Coutinho, F.; Cortesão, R. Online Stiffness Estimation for Robotic Tasks with Force Observers. *Control. Eng. Pract.* **2014**, *24*, 92–105. [CrossRef]
19. Hattori, G.; Serpa, A.L. Contact Stiffness Estimation in Ansys Using Simplified Models and Artificial Neural Networks. *Finite Elem. Anal. Des.* **2015**, *97*, 43–53. [CrossRef]
20. Gao, Y.; Hendricks, L.A.; Kuchenbecker, K.J.; Darrell, T. Deep Learning for Tactile Understanding from Visual and Haptic Data. In Proceedings of the 2016 IEEE International Conference on Robotics and Automation (ICRA), Stockholm, Sweden, 16–21 May 2016; pp. 536–543.
21. Wettels, N.; Santos, V.; Johansson, R.; Loeb, G. Biomimetic Tactile Sensor Array. *Adv. Robot.* **2008**, *22*, 829–849. [CrossRef]

22. Chin, K.; Hellebrekers, T.; Majidi, C. Machine Learning for Soft Robotic Sensing and Control. *Adv. Intell. Syst.* **2020**, *2*, 1900171. [CrossRef]
23. Zimmer, J.; Hellebrekers, T.; Asfour, T.; Majidi, C.; Kroemer, O. Predicting Grasp Success with a Soft Sensing Skin and Shape-Memory Actuated Gripper. In Proceedings of the 2019 IEEE/RSJ International Conference on Intelligent Robots and Systems (IROS), Macau, China, 4–8 November 2019; pp. 7120–7127. [CrossRef]
24. Al-Ibadi, A.; Nefti-Meziani, S.; Davis, S. Controlling of Pneumatic Muscle Actuator Systems by Parallel Structure of Neural Network and Proportional Controllers (PNNP). *Front. Robot. AI* **2020**, *7*, 115. [CrossRef]
25. Thuruthel, T.G.; Falotico, E.; Renda, F.; Laschi, C. Model-Based Reinforcement Learning for Closed-Loop Dynamic Control of Soft Robotic Manipulators. *IEEE Trans. Robot.* **2019**, *35*, 124–134. [CrossRef]
26. Bern, J.M.; Schnider, Y.; Banzet, P.; Kumar, N.; Coros, S. Soft Robot Control With a Learned Differentiable Model. In Proceedings of the 2020 3rd IEEE International Conference on Soft Robotics (RoboSoft), New Haven, CT, USA, 15 May–15 July 2020; pp. 417–423.
27. Rotella, N.; Mason, S.; Schaal, S.; Righetti, L. Inertial Sensor-Based Humanoid Joint State Estimation. *arXiv* **2016**, arXiv:1602.05134v1.
28. Ancillao, A.; Tedesco, S.; Barton, J.; O'Flynn, B. Indirect Measurement of Ground Reaction Forces and Moments by Means of Wearable Inertial Sensors: A Systematic Review. *Sensors* **2018**, *18*, 2564. [CrossRef]
29. Culbertson, H.; Delgado, J.J.L.; Kuchenbecker, K.J. *The Penn Haptic Texture Toolkit for Modeling, Rendering, and Evaluating Haptic Virtual Textures*; ResearchGate: Berlin, Germany, 2014.
30. Slobodenyuk, N.; Jraissati, Y.; Kanso, A.; Ghanem, L.; Elhajj, I. Cross-Modal Associations Between Color and Haptics. *Atten. Percept. Psychophys.* **2015**, *68*, 1379–1395. [CrossRef]
31. Ji, M.; Fang, L.; Zheng, H.; Strese, M.; Steinbach, E. Preprocessing-free Surface Material Classification using Convolutional Neural Networks Pretrained by Sparse Autoencoder. In Proceedings of the IEEE International Workshop on Machine Learning for Signal Processing, Boston, MA, USA, 17–20 September 2015.
32. Tomovic, R.; Boni, G. An Adaptive Artificial Hand. *Autom. Control. Ire Trans.* **1962**, *7*, 3–10. [CrossRef]
33. Aukes, D.; Heyneman, B.; Ulmen, J.; Stuart, H.; Cutkosky, M.; Kim, S.; Garcia, P.; Edsinger, A. Design and testing of a selectively compliant underactuated hand. *Int. Robot. Res.* **2014**, *33*, 721–735. [CrossRef]
34. Mańkowski, T.; Tomczyński, J.; Walas, K.; Belter, D. PUT-Hand—Hybrid Industrial and Biomimetic Gripper for Elastic Object Manipulation. *Electronics* **2020**, *9*, 1147. [CrossRef]
35. You, W.S.; Lee, Y.H.; Oh, H.S.; Kang, G.; Choi, H.R. Design of a 3D-printable, robust anthropomorphic robot hand including intermetacarpal joints. *Intell. Serv. Robot.* **2019**, *12*, 1–16. [CrossRef]
36. Kang, L.; Seo, J.T.; Kim, S.H.; Kim, W.J.; Yi, B.J. Design and Implementation of a Multi-Function Gripper for Grasping General Objects. *Appl. Sci.* **2019**, *9*, 5266. [CrossRef]
37. Foumani, M.; Gunawan, I.; Smith-Miles, K.; Ibrahim, M.Y. Notes on Feasibility and Optimality Conditions of Small-Scale Multifunction Robotic Cell Scheduling Problems With Pickup Restrictions. *IEEE Trans. Ind. Inform.* **2015**, *11*, 821–829. [CrossRef]
38. Shintake, J.; Cacucciolo, V.; Floreano, D.; Shea, H. Soft Robotic Grippers. *Adv. Mater.* **2018**, *30*, 1707035. [CrossRef] [PubMed]
39. Suzumori, K.; Iikura, S.; Tanaka, H. Applying a Flexible Microactuator to Robotic Mechanisms. *IEEE Control. Syst. Mag.* **1992**, *12*, 21–27. [CrossRef]
40. Terryn, S.; Brancart, J.; Lefeber, D.; Van Assche, G.; Vanderborght, B. Self-healing soft pneumatic robots. *Sci. Robot.* **2017**, *2*. [CrossRef]
41. Gu, G.Y.; Zhu, J.; Zhu, L.M.; Zhu, X. A Survey on Dielectric Elastomer Actuators for Soft Robots. *Bioinspir. Biomimetics* **2017**, *12*, 011003. [CrossRef]
42. Sreekumar, M.; Nagarajan, T.; Singaperumal, M.; Zoppi, M.; Molfino, R. Critical Review of Current Trends in Shape Memory Alloy Actuators for Intelligent Robots. *Ind. Robot. Int. J.* **2007**, *34*, 285–294. [CrossRef]
43. Liu, C.H.; Huang, G.F.; Chiu, C.H.; Pai, T.Y. Topology Synthesis and Optimal Design of an Adaptive Compliant Gripper to Maximize Output Displacement. *J. Intell. Robot. Syst.* **2018**, *90*, 287–304. [CrossRef]
44. Xu, Z.; Todorov, E. Design of a Highly Biomimetic Anthropomorphic Robotic Hand Towards Artificial Limb Regeneration. In Proceedings of the 2016 IEEE International Conference on Robotics and Automation (ICRA), Stockholm, Sweden, 16–21 May 2016; pp. 3485–3492. [CrossRef]
45. Walker, J.; Zidek, T.; Harbel, C.; Yoon, S.; Strickland, F.S.; Kumar, S.; Shin, M. Soft Robotics: A Review of Recent Developments of Pneumatic Soft Actuators. *Actuators* **2020**, *9*, 3. [CrossRef]
46. Hochreiter, S.; Schmidhuber, J. Long Short-Term Memory. *Neural Comput.* **1997**, *9*, 1735–1780. [CrossRef] [PubMed]
47. Schuster, M.; Paliwal, K. Bidirectional Recurrent Neural Networks. *IEEE Trans. Signal Process.* **1997**, *45*, 2673–2681. [CrossRef]

Article

PUT-Hand—Hybrid Industrial and Biomimetic Gripper for Elastic Object Manipulation

Tomasz Mańkowski *, Jakub Tomczyński, Krzysztof Walas and Dominik Belter

Institute of Robotics and Machine Intelligence, Poznań University of Technology, ul. Piotrowo 3A, 60-965 Poznań, Poland; jakub.tomczynski@put.poznan.pl (J.T.); krzysztof.walas@put.poznan.pl (K.W.); dominik.belter@put.poznan.pl (D.B.)
* Correspondence: tomasz.mankowski@put.poznan.pl

Received: 16 June 2020; Accepted: 10 July 2020; Published: 16 July 2020

Abstract: In this article, the design of a five-fingered anthropomorphic gripper is presented specifically designed for the manipulation of elastic objects. The manipulator features a hybrid design, being equipped with three fully actuated fingers for precise manipulation, and two underactuated, tendon-driven digits for secure power grasping. For ease of reproducibility, the design uses as many off-the-shelf and 3D-printed components as possible. The on-board controller circuit and firmware are also presented. The design includes resistive position and angle sensors in each joint, resulting in full joint observability. The controller has a position-based controller integrated, along with USB communication protocol, enabling gripper state reporting and direct motor control from a PC. A high-level driver operating as a Robot Operating System node is also provided. All drives and circuitry of the PUT-Hand are integrated within the hand itself. The sensory system of the hand includes tri-axial optical force sensors placed on fully actuated fingers' fingertips for reaction force measurement. A set of experiments is provided to present the motion and perception capabilities of the gripper. All design files and source codes are available online under CC BY-NC 4.0 and MIT licenses.

Keywords: robotic hand; control; perception; tactile sensing; mechatronics; grasping; manipulation; PUT-Hand; underactuated

1. Introduction

Manipulation of elastic pipes and wires in factory environments is performed mainly by human operators. Some of the tasks on the elastic objects are performed by the machines designed specifically for that operation. Rarely, the manipulation of elastic objects is performed by robots which can adapt to the changes in the process. This scenario is still challenging, due to the deficiencies in mechanical design of the grippers and the perception systems. Interaction with elastic objects requires not only high manoeuvrability but also reliable tactile feedback.

Robots have a potential to manipulate elastic objects autonomously without a supervision of a human operator. Application of robots in factories improves the production process and reduces number of errors and mistakes made by humans. However, manoeuvrability of the robot and capability to work with elastic objects still need development.

Capabilities of the grippers available in the industry are limited, as most of them are two-fingered or three-fingered [1–3]. They are often fully actuated systems with position-based control, designed for precise manipulation. This means that they are not designed to deal with variability and uncertainty of an environment, a slight change in object positioning or its shape may cause a failure.

Contrarily, many adaptive grippers are designed to replace missing body parts, mimicking the human hand, with five fingers [4]. Digits can be both fully actuated [5] or underactuated for better adaptation to shape of grasped objects [6]. However, they are designed to replace human hands, not to operate as a gripper of an industrial robot.

In this research, we focus on the design of a robotic hand which can be used to manipulate elastic objects. We carefully study the literature to find a compromise between precision of the grasp and adaptability, achieved by optimising the number of fully actuated and underactuated fingers. We also study the perception system of robotic hands and propose the application of optical tactile sensors on the fingertips. We present the application of the hand in several tasks related to manipulation of elastic objects, including grasping, tactile sensing, and application-oriented tasks.

1.1. Problem Statement

Our main goal was to design a compact robotic hand which has the capability to interact with elastic objects. The hand should enable precise manipulation and grasping, while allowing power grasping with underactuated fingers. We had to determine which joints of the fingers should be fully actuated and which ones should be underactuated. Moreover, we had to design the actuation mechanisms to preserve compact dimensions of the hand.

The design of the hand should provide full observability of all joints, including underactuated ones, as the state of the joint does not depend only on the drive, but also on the shape of grasped object, and the state of neighbouring joints. The perception system of the hand should also allow measuring mechanical properties of objects and reaction forces.

Finally, the mechanical design and perception system should allow performing task-specific modelling. During interacting with elastic objects the sensors should provide information about the contact with the object and measure the changes in the object's state. The example scenario is plug insertion, during which reaction forces increase until the plug reaches a stable configuration. The final state of the system should be detected by the sensory system of the proposed hand.

1.2. Approach and Contribution

In this article, we present a new open-source design of the dexterous robotic hand - PUT-Hand, shown in Figure 1. The gripper includes fully actuated fingers with tactile sensing, and underactuated fingers for power and adaptive grasps. The mechanical part is providing in one design the compact human-like hand for skilful manipulation with hybrid fully actuated and underactuated fingers. The hand is lightweight and has a unique feature that all the drives are built-in the palm of the hand, which is reducing the length of the end effector hence it is minimally shrinking work volume of manipulator equipped with our hand. The hand is reproducible. The mechanical and PCB design files, firmware, and ROS node are available at https://github.com/puthand under CC BY-NC 4.0 and MIT licenses. We also provide a broad literature review related to robotic and prosthetic hand designs. We also contribute to the control and perception system of the hand, presenting hand on-board controller and a ROS driver. The ROS driver enables computing forward and inverse kinematics, motion planning, and visualisation. The full software stack is available.

The hand is designed for elastic object manipulation by providing three fingered-like manipulation when thumb in the first joint is oriented towards the inner part of the palm. The hand operates in robotic gripper mode and it allows in hand dexterous manipulation of elastic elongated objects i.e., cables and hoses. Additionally, thanks to full sensorisation of the hand, joint encoders and force sensors at the fingertips, one can estimate elastic object parameters by performing in hand manipulation and measuring fingers displacement and reaction forces caused by the movement of the fingers.

Finally, we provide an experimental verification of PUT-Hand and its controllers. We show manoeuvrability of PUT-Hand during grasping various objects. We demonstrate that the sensory system of the hand can be used to detect contact with an object but also to measure its physical properties (e.g., object stiffness). We also provide application-oriented experiments to show the interaction of the hand with elastic objects during pipe bending and inserting a plug into a socket.

Figure 1. PUT-Hand—An open source dexterous robotic hand.

2. Related Work

We limit the review to hand designs available in the literature, also considering commercially available hands only if description and specification are detailed enough. For this reason, we do not include some of the commercial prosthetic hands, such as VINCENT hands by Vincent Systems [4], iLimb and iLimb Pulse by Touch Bionics [7], and Michelangelo by Ottobock [8]. If a research team publishes a series of designs, we only consider the latest version in the review. Multiple hand designs by the same group are only considered if the mechanical concept of the hands differs significantly.

Most of the hands considered in the review are research platforms presenting unique mechanical, control or sensing concepts. Additionally, we include five commercial platforms in the comparison. In the review, we have proposed gripper design taxonomy based on three criteria: joint drive mechanism type, whether the system is fully actuated or not, and number of digits and kinematic structure.

2.1. Joint Drive Mechanisms

A large variety of drive to joint power transfer methods can be found in the literature. In this work we analyse the most common and representative approaches. The classification of joint drive mechanisms for robotic fingers is presented in Figure 2 as eight categories: direct drive with gears, fully actuated joints with tendons, backdrivable tendon, underactuated finger with a tendon, rigid linkage system, underactuated joints with rods, pneumatically actuated, and pneumatic/hydraulic muscles. We identified 21 structures with tendon mechanisms [2,3,6,9–26], two with direct drive based on gears [1,5], one rod-based [27], two with rigid linkage system [28,29], one with differential drivetrain and cams [17], one with pneumatic chamber [30], and one with hydraulic muscles [31]. All the cited papers were grouped in Table 1.

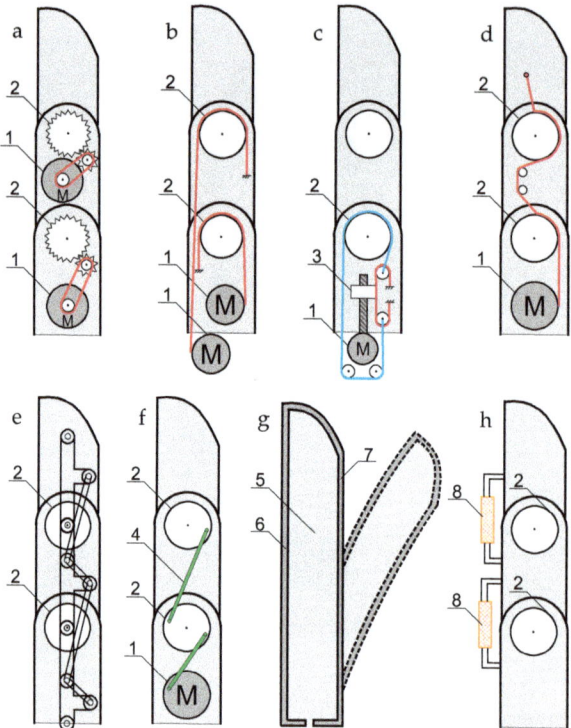

Figure 2. Joint drive mechanisms for robotic fingers: direct drive with gears (**a**), fully actuated joints with tendons (**b**), backdrivable tendon (**c**), underactuated finger with tendon (red or blue) (**d**), rigid linkage system (**e**), underactuated joints with rods (green) (**f**), pneumatically actuated (**g**), pneumatic/hydraulic muscles (**h**). Legend: 1—electrical motor, 2—joint, 3—nut, 4—rod, 5—air chamber, 6—elastic fabric, 7—inelastic fabric, 8—pneumatic/hydraulic muscles.

Table 1. Related work—joint drive mechanisms review.

Tendon mechanism	[2,3,6,9–26]
Gears	[1,5]
Rod-based	[27]
Rigid linkage system	[28,29]
Differential drivetrain	[17]
Pneumatic chamber	[30]
Hydraulic muscles	[31]

The most popular method for joint coupling uses tendons (Figure 2b). The tendon transfers the torque from an electrical motor or force from a linear actuator to the rotational joint. The tendon might be coupled with a spring to generate compliant behaviour of the fingers. The popularity of tendon-like mechanisms comes from the fact that this mechanism is bio-inspired and gives the natural compliance and adaptation of the hand. This approach reduces the number of actuators and allows obtaining a robust grasp without additional tactile or force/torque sensors.

Gears are used mainly in the fully actuated fingers to transfer the torque from the drive to the joint [9,11,31]. Rods become popular when the drives are located in the palm and the rods push the finger links [27]. This approach reduces the weight of the fingers. The lighter hands are actuated by the pneumatic chambers [30]. In this case the fingers are made of elastic material and bend when the pressure inside the chamber changes. Moreover, the finger adapts to the shape of the object [30]. On the other hand, the gripper with pneumatic chambers requires the air compressor. The same problem exists when pneumatic [22] and hydraulic [31] muscles are used.

The fingers of PUT-Hand for the first three fingers are fully actuated with rigid linkage system and gears [28,29]. The last two fingers are tendon driven, e.g., [15].

2.2. Mechanical and Kinematic Structure

In the review we identified six grippers which are fully actuated [1,9,11,22,31,32]. However, most of the designs exploit the underactuation concept: [2,3,6,14–16,18–21,23–25,27–30,33], ref. [5,10] (two joints coupled), ref. [12] (four joints coupled), ref. [13] (20 joints/9 actuated), ref. [17] (18 joints/16 actuated). Furthermore, the review revealed the following categories with respect to digit count: three-fingered [1–3], four-fingered [6,9,11,25,34]. Most of the grippers described in the literature have five fingers: [5,10,12–17,17–24,27–32]. All the hand structures were grouped in Table 2 regarding mechanical structure and in Table 3 regarding kinematic structure.

Table 2. Related work—Robotic hands mechanical structure review.

Fully actuated	[1,9,11,22,31,32]
Underactuated	[2,3,6,14–16,18–21,23–25,27–30,33]
Hybrid designs	[34–41]

Table 3. Related work—Robotic hands kinematic structure review.

Three-fingered	[1–3]
Four-fingered	[6,9,11,25,34,39,40]
Five-fingered	[5,10,12–17,17–24,27–32,35]

In our design we decided to use five fingered design closely resembling human hand in geometry and size as it was outlined in [35]. This choice allows us to perform dexterous manipulation and different grasps and manipulation strategies are obtained in software without changes in the mechanical hardware of the hand.

2.2.1. Fully Actuated Designs

Most of commercially available grippers have a limited number of joints, with full actuation and position sensing (e.g., 3-finger Schunk SDH [1]). In a backdrivable CEA hand [10] the joints are fully actuated, with last two joints in each digit mechanically coupled, providing natural behaviour of the hand during force control and high robustness. Utah/M.I.T Dextrous hand [9] is a 4-fingered tendon-driven design, where 32 drives are required. Another 4-fingered design, Sandia hand, was designed by Quigley et al. [11], with each joint actuated by brushless DC servomotors and tendons. Focus was put on the robustness of the hand, and fingers separate from the hand in case of a collision. The UB Hand 3 [17] has 16 degrees of mobility, with joints driven by tendons.

Grasping control algorithm problems become apparent in more complex fully actuated 5-fingered designs [5]. The gripper with the most complex kinematic structure which is considered in this review is Shadow hand [12]. The hand has 5 fingers and 24 joints actuated by McKibben muscles and tendons, which makes it robust to disturbances, but also mechanically complex and of substantial size and weight. The latest hand built at the University of Washington has a similar kinematic structure, with two versions built: driven by pneumatic cylinders [22] and driven by servomotors [23]. Both versions use tendons to transfer the energy from drives to the joints and the drives are mounted in the forearm.

When using a fully actuated design, all joints angles have to be computed according to given dimensions of handled object. If the measurements are uncertain the hand has limited capabilities to adapt to the shape, or the adaptation is slow due to delayed feedback from sensors. On the other hand, fully actuated grippers can be used to perform more complex manipulation tasks. The digits can reach any position in the workspace in contrast to underactuated systems, where motion trajectory also depends on the shape of the manipulated object.

2.2.2. Underactuated Designs

Underactuation concept not only allows designing compliant hands which adapt to the shape of the objects but also to reduce the number of actuators and weight of the hand. One of the first underactuated anthropomorphic grippers was built at the University of California and Belgrade [14]. The hand has 5 fingers and 14 joints, driven by two servomotors and a tendon-driven mechanism. The elastic coupling of the joints allows the fingers to adjust to the shape of an object. TBM hand [28] is a five-finger design using a rigid linkage system to couple joints, all joints are driven by a single DC servomotor. Compliance is obtained by application of extension springs pulling the linkage systems. Similarly, ref. [20] is an underactuated structure driven by one DC motor.

Four fingered design is described in [25] the project uses simple 3D-printed components with compliant flexure joints and off-the shelf parts to provide low-cost, open-source underactuated hand.

Three-fingered designs often mimic the behaviour of the thumb, index, and middle finger of the human hand. An underactuated prosthesis proposed by Zollo et al. [2] is optimised to be capable of reproducing natural human motions. Similarly, the 3-fingered RTR II hand [3] uses differential mechanisms to control multiple joints. The BarretHand designed for industry [42] uses the mechanism which shifts torque to appropriate finger joint. The joints are locked when the fingertip reaches contact with the object. The number of actuators is limited in SRI hand [6] and two fingers can be rotated on a slider mechanism. Breaks in joints are used to reduce power consumption and motor torque during a steady grasp. The same authors also propose the tool that is aimed to design underactuated hands [43]. The simulation tool takes into account the dynamics of the hand, an actuation mechanism, and contact friction. The core of the simulation engine is based on three-dimensional force fields. Southampton-Remedi hand [29] has six drivable degrees of freedom. The authors of [24,44] explore the concept of active synergies in performing grasping and manipulation tasks. The compliance of SmartHand [18,19] is obtained using Hirose's soft (differential) finger mechanism [45]. Various hands were proposed by groups which are interested in anthropomorphic prosthetics. An underactuated MANUS hand [16] has five fingers but only three of them are actuated. The hand has four grasping modes: cylindrical, precision, hook and lateral obtained with two servomotors only. Due to simplicity in mechanical design underactuated grippers can be easily prototyped using 3D printing [27,46].

Underactuated designs have a reduced controllability compared to fully actuated ones; however, the general trend is to reduce the number of actuators. The Vanderbilt hand [21] has five fingers and 16 joints driven by tendons and five DC motors. The UNB anthropomorphic hand [33] has five fingers and only three DC motors. The complex motions are obtained using differential drivetrain and cams.

To achieve required dexterity while keeping the design compact and control algorithms simple, hybrid designs combining both approaches can be explored, with fully actuated fingers dedicated to precise manipulation and underactuated supporting fingers for power grasp. One of the possible solutions was presented in [34], where authors presented a unique design of the fingers which allows for generating linked and adjustable motions. Joints exhibit a coupled movement in free space and moves adaptively when in contact with the objects. Word hybrid applied to hand-design is understood differently in the literature. For Mizushima et al. hybrid was used in the context of hybrid design of the fingers, which skeleton was tendon driven and inside the skeleton the granular material after removing the air the grasping posture can be fixed [36]. A similar understanding of word hybrid is present in [38]. In the case of work presented, in [37] word hybrid is used for describing the hand with three mechanical and three soft fingers. Additionally, in work [35] word hybrid was used to characterise linkage and tendon driven-based fingers. In work done by Jeong and Cheong, the hybrid nature of the hand is understood as the mode of operation. Hand uses four fingers for human-like motions in human hand mode, and three fingers without the thumb when it is used in conventional robotic hand mode [39,40]. Conversely, Cerruti et al. use word hybrid to describe two actuation system present in hand and working in parallel. One is responsible for gesture capability but low force (linkage mechanism) and the second is used when the grasping force is needed (tendons) [41].

In the case of PUT-Hand, the hybrid is understood two folds. First, the hand is composed of three actuated fingers and two underactuated. Second, PUT-Hand, like the design proposed in [39,40], has two modes of operation, which in case of a PUT-Hand depends on the orientation of the thumb. When the thumb in the first joint is oriented towards the inner part of the palm the hand operates in robotic gripper mode similar to BarretHand [42]. This configuration allows the robot to perform in hand manipulation of the elastic objects such as cables. When the thumb is rotated outwards from the palm, it resembles a human hand as in [35].

2.3. Tactile Sensing

A suitable sensory system is crucial while performing successful grasping and manipulation tasks, particularly for elastic object handling. In autonomous, unsupervised manipulation, initial gripper configuration can be chosen based on an object model obtained by a RGB-D sensor [47,48]. Vision-based grasping algorithms can be further improved by active-sensing—a strategy for view selection to maximise the surface reconstruction and safety of the planned trajectory [49]. However, data from depth camera can be noisy, causing uncertainty in generated model [50], leading to a decrease in overall autonomous manipulation performance or even a failure.

Equipping a gripper with tactile sensors can further improve its manipulation capabilities in robotised setups by providing feedback to the robot controller [51], estimating the contact force and actively controlling the reaction forces to stabilise the grasp [52], or as a source of object identification [53,54]. Tactile information can also be used to determine physical properties of an elastic object or to determine the state of manipulated object. Tomo et al. [55] has proposed uSkin 3D tactile sensors intended for Allegro Hand fingertips and phalanges, providing 16 independent force measurements for each contact plane. Data from these sensors can be used as an input for Convolutional Neural Networks in tasks of object identification [56].

Other designs introduce tactile sensing using: inertial units [57], piezoelectric sensors [58,59], resistive sensors [60], a camera for deformation measurement [61], or capacitive sensors [62,63]. An extended literature review in the field of touch sensing is presented in [64,65].

3. PUT-Hand Design

3.1. Mechanical Design

The main design goal of PUT-Hand project was to create an anthropomorphic gripper which is capable of performing both precise object manipulation and power grasps, in an industrial environment, while using as many off-the-shelf parts as possible. A taxonomy of human grasps [66] shows that most of precision grasps and manipulations is done using three or less fingers—thumb, index, and middle. Many grasps use virtual fingers, where several fingers work as one functional unit. Ring and little fingers are used mostly as assisting fingers in power grasps, where individual control of each joint is not necessary. To provide a balance between grasping capabilities and complexity of the gripper's mechanical and control structures, a hybrid structure was proposed. The design incorporates three fully actuated fingers (thumb, index, and middle), and two underactuated fingers. All movable joints, in both actuated and underactuated fingers, are fitted with position sensors and observable, allowing for full grasping planning and simulation. The resulting design, shown in Figure 3, is mostly 3D-printable, with single elements requiring CNC machining from aluminium or turning from stainless steel. All drives and controllers are integrated within the palm or fingers.

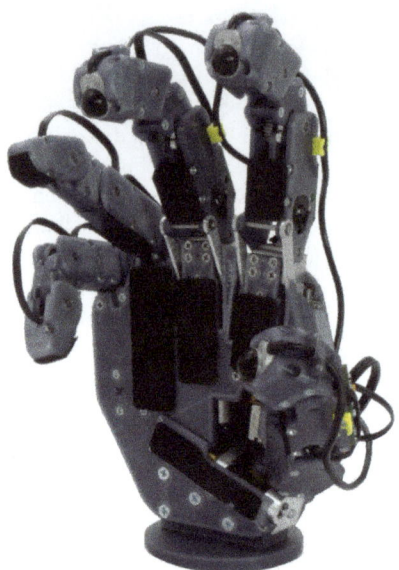

Figure 3. PUT-Hand—an open-source dexterous robotic hand.

3.1.1. Fully Actuated Fingers (Index and Middle)

Index and middle fingers are designed as fully actuated, with two degrees of freedom. MCP is driven independently, while PIP and DIP share the same drive. This configuration provides full control over fingertip position in finger's 2D plane while maintaining mechanical simplicity (MCP is a single DoF joint with no adduction or abduction ability). The fingers share the same design and dimensions, allowing for easier manufacturing and parts interchangeability. Overall finger dimensions of 18 mm (width) by 20 mm (height) closely correlate with adult mean index finger width [67]. Annotated design of the finger is shown in Figure 4.

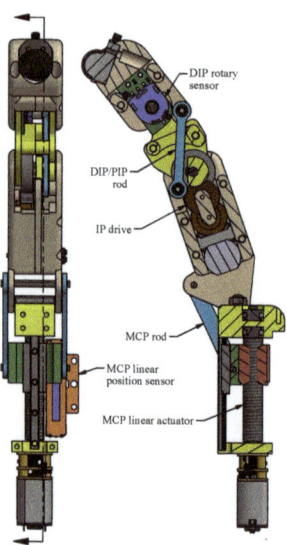

Figure 4. Mechanical design and drives of a fully actuated finger; design intended for index and middle fingers.

Due to large torque requirements in MCP, a linear actuator was designed to drive the joint. The drive uses a Pololu Micro High Power 6 V DC motor with 1:75 gearbox driving an Igus DryLin® lead screw (2.54 mm pitch). The lead screw is supported by a set of two thrust bearings on the opposite side. The lead nut is attached to a feed sliding on a CPC MR3ML ball bearing linear guide. Drive position feedback is obtained from an Alps RDC1022A05 linear resistive position sensor.

The linear actuators fit within the palm of the gripper and flex the fingers using aluminium rods. The relation between actuator position and finger flexion is shown in Figure 5. The structure reaches maximum efficiency at $\theta_{MCP} = 61°$, exerting the force of 8.7 N at the fingertip.

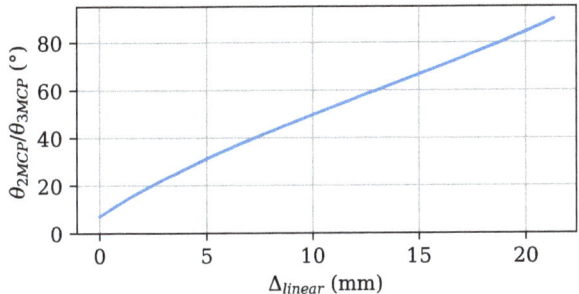

Figure 5. Relation between linear actuator movement and finger flexion in MCP.

The second drive resides inside the proximal phalanx. The drive uses motor and metal gears from Hitec (gear set 56,396), used in a range of miniature digital servos (e.g., HS-5245MG). The gearbox was reconfigured to fit in narrow space of the phalanx and drives PIP directly. Movement is passed to DIP using an aluminium rod. Angle feedback is provided at DIP using an Alps RDC503013A rotary resistive sensor. Relation between PIP and DIP angles is shown in Figure 6. The rotation ratio between PIP and DIP was chosen, so with both drives fully flexed, the fingertip touches the metacarpus. When fully extended, the drive exerts the force of 8.0 N at the fingertip.

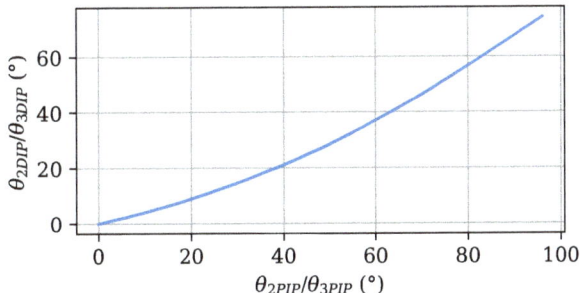

Figure 6. Relation between flexion in PIP and DIP.

The finger features an interchangeable fingertip, allowing for mounting of a 3D OptoForce OMD-10 force transducer (shown in the drawings) or a resin moulded passive fingertip, according to particular task requirements.

3.1.2. Thumb

Inclusion of an opposing thumb is vital to the performance of many types of grasps. The thumb has three degrees of freedom—two joints in planar configuration (MCP and IP), and third CMC placed at an angle. Thumb's MCP and IP drive share a similar design to index and middle fingers' PIP. The DC motors and gearing are the same, but gearbox layout was altered to fit in the available space in metacarpal and proximal phalanx. Thumb design is shown in Figure 7.

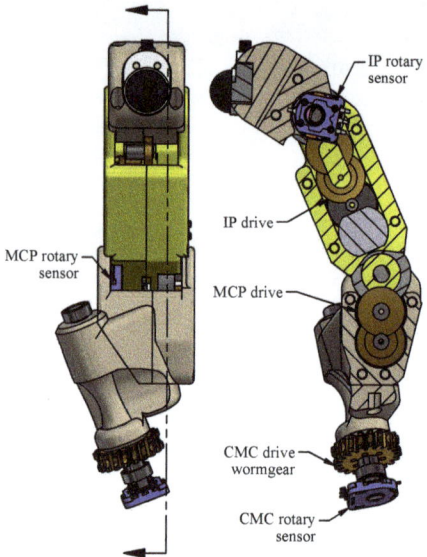

Figure 7. Mechanical design and drives of fully actuated thumb; CMC drive motor and worm (not visible) are enclosed in the metacarpus.

Due to the nature of CMC operation, which is moved mainly when switching between various grasping types, a worm gear drive was used. The drive uses a Pololu Micro High Power 6 V DC motor with 1:298 gearbox with an additional 1:20 reduction provided by the worm drive. Resistive rotary sensors by the Alps were used to provide angle feedback from all joints.

3.1.3. Underactuated Finger

Ring and little fingers were designed as underactuated structures to simplify both mechanical design and control algorithms. The fingers use a tendon flexion mechanism, as seen in Figure 2d. The tendon is made from a braided fishing line, providing low extensibility, while maintaining high compliance. Each tendon is wound onto a spool driven by a Pololu Micro High Power 6 V DC motor with a 1:298 gearbox.

An eccentric ring is placed around each joint axle and attached to a pair of helical extension springs. While in other approaches springs have been used to store potential energy for efficiency improvements [68], here they are required to return the finger to its extended position and distribute flexion evenly among the three joints. To limit the number of required off-the-shelf part types, all joints share the same type of springs, with only the anchor radius of the spring differentiating the extension torque. The torque was adjusted experimentally to allow the fingers to straighten in the most demanding orientation (with palm facing downwards), without adding too much additional resistance to the drive. The overall design of the finger is shown in Figure 8. Both fingers share the same design, but proximal and intermediate phalanges are shorter in the little finger.

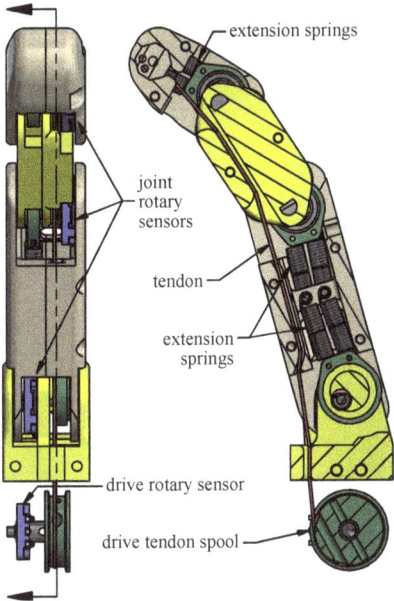

Figure 8. Mechanical design of an underactuated finger; design intended for ring and small fingers; drive motor not visible.

To provide full joint state feedback to motion planning software, each joint is equipped with a resistive rotary sensor. An additional rotary sensor is mounted at the spool shaft to enable proper drive control. To summarise the mechanical concept of the PUT-Hand a short comparison with typical robotic hand is provided. Namely, DLR-HIT Hand II [5] and Pisa/IIT Softhand 2 [24]. DLR-HIT Hand II is fully actuated with 15 degrees of freedom, three in each finger. There are five identical fingers. All the drives are embedded in the palm. The motion of each finger is controllable. On the contrary, the PISA IIT Softhand 2 has 19 anthropomorphic degrees of freedom controlled by one motor and relies fully on underactuation concept. PUT-Hand exploits the best of two worlds. It is fully actuated, as DLR-HIT Hand II, for the first three fingers. These fingers are needed for precise manipulation and force sensing. Additionally, it is exploiting underactuation, as Pisa/IIT Softhand 2, for the last two fingers which are supporting the grasping of larger objects.

3.2. Controller

The low-level controller of PUT-Hand was designed with high modularity in mind. The main unit, called HUB, is responsible for communication with the high-level controller (e.g., a PC), communication with individual servomechanisms, and (optionally) internal drive position control. Each servomechanism (referred to as DRIVER) is a separate, independent module consisting of a direct-current (DC) motor, an adequate number of encoders, and a printed circuit board with all necessary communication and motor control circuitry. Overall controller architecture is shown in Figure 9. DRIVER modules are connected to the HUB using a bus configured in a star pattern, which allows for easy replacement of faulty drives integrated into the mechanical design of the hand. Controller attachment and arrangement on PUT-Hand is visible in Figure 10. The system is fully scalable and can be used in designs with various drive configurations, not limited to grippers.

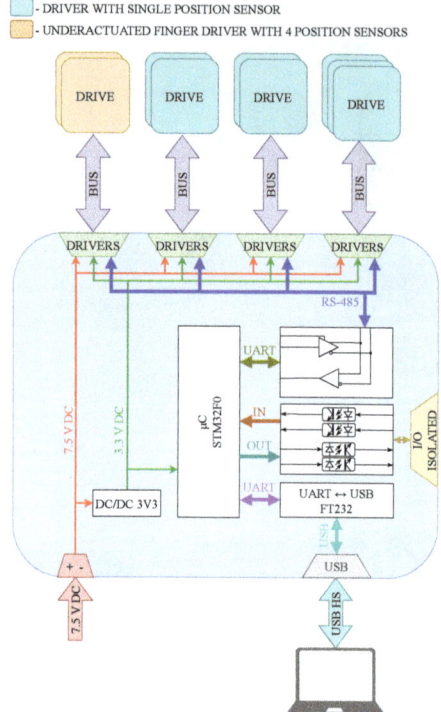

Figure 9. PUT-Hand low-level controller architecture, with DRIVER modules connected in star pattern allowing easy interchangeability of components; main unit (HUB) diagram showing circuit components.

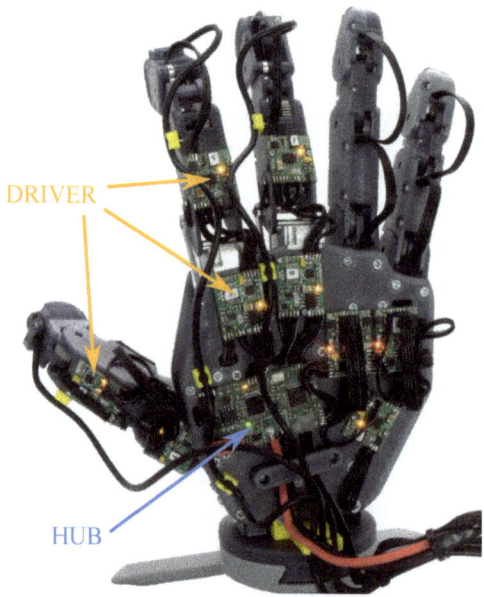

Figure 10. PUT-Hand with visible controller electronics; not all DRIVER modules are marked.

3.2.1. HUB Design

HUB is an integration unit serving as a bridge between each separate DRIVER and high-level controller. Detailed architecture of the HUB is presented in Figure 9. The unit is controlled by STM32F0 family microcontroller with ARM Cortex-M0 core running at 48 MHz. Communication with PC is carried out using a full-duplex universal asynchronous serial bus (UART), for the ease of use, a UART $\Longleftrightarrow$ USB converter by FTDI was embedded. The system is powered using 7.5 V DC, HUB includes a DC/DC step-down converter to 3.3 V DC. Both supply voltages are distributed to populate DRIVER connectors. Communication with DRIVER modules is implemented using a half-duplex RS-485 hardware protocol, with a switchable transceiver. Moreover, HUB unit implements two pairs of isolated input and output for additional high-level control purposes, for operating modes without a PC.

3.2.2. DRIVER Design

Each DRIVER is a separate servomechanism unit consisting a printed circuit board, a DC motor and resistive position sensors, presented in Figure 11. Two types, with the different number of position sensor connectors, are used in the system. Standard DRIVER supports only one potentiometer, in case of underactuated finger configuration a board with support for four sensors is used. The board features a RS-485 transceiver in half-duplex mode with the a switchable driver. A Texas Instruments DRV8872 MOSFET-based H-bridge with peak 3.6 A current capacity serves as an execution circuit in controlling brushed DC motor. An integrated H-bridge with over-current, under-voltage, and over-temperature protection, including fault interrupt pin is used. High-current side of the motor driver is supplied by 7.5 V DC. Remaining parts of the DRIVER board are supplied by 3.3 V DC. DRIVER also implements a motor current measurement circuit based on a sense resistor. Acquisition of all sensor data, PWM generation, and communication with the HUB unit are carried out by low-power STM32L0 family microcontroller with ARM Cortex-M0+ core running at 32 MHz.

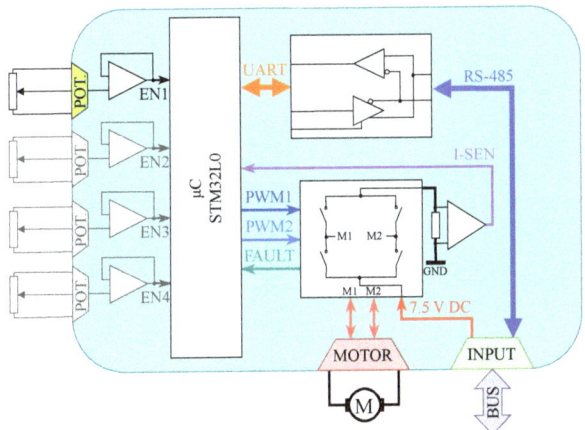

Figure 11. Diagram of DRIVER servo unit; two configurations, with one or four resistive position sensors, are used in the system.

3.3. Firmware

DRIVER is only an execution unit, and collects sensor data about itself (position, motor current), it does not perform any position control tasks of the drive. Using a defined protocol, secured with a CRC-8 checksum, HUB unit reads the current status of each particular DRIVER and sets the PWM duty for each DC motor, together with the rotation direction. DRIVER modules are addressed using a unique address stored in μC EEPROM memory. In case of a communication loss of over 50 ms, the DRIVER will engage an electronic motor brake. The protocol uses normalised values for all communication, and all individual drive data is stored internally. This way, a DRIVER can be easily replaced in case of a failure, even with a drive of different type, while remaining transparent to the HUB and overall control scheme.

HUB serves as drive control unit, performing cyclic communication with all DRIVERs at the frequency of 100 Hz. It allows for control of drives in 3 primary modes:

- *Idle*—where all DRIVERs engage electronic brake or disable the H-bridge, depending on user's choice.
- *Internal*—where HUB's internal PID controller with dead-zone is used to position fingers. In this mode, user sets a desired fingers position via the USB interface. Internal PID does not provide force regulation, motor currents are neglected. A diagram of internal control mode is presented in Figure 12a.
- *External*—in this mode HUB acts as a middleman between external user implemented controller and particular drives, providing information about DRIVERs status and forwarding PWM duty. A diagram of external control mode is presented in Figure 12b.

Most communication with PC (high-level controller) is performed on PC request; however, a cyclic status report can be enabled. In this case, the HUB will transmit a full data vector describing hand status (positions, motor currents, mode, etc.) with a frequency of 100 Hz.

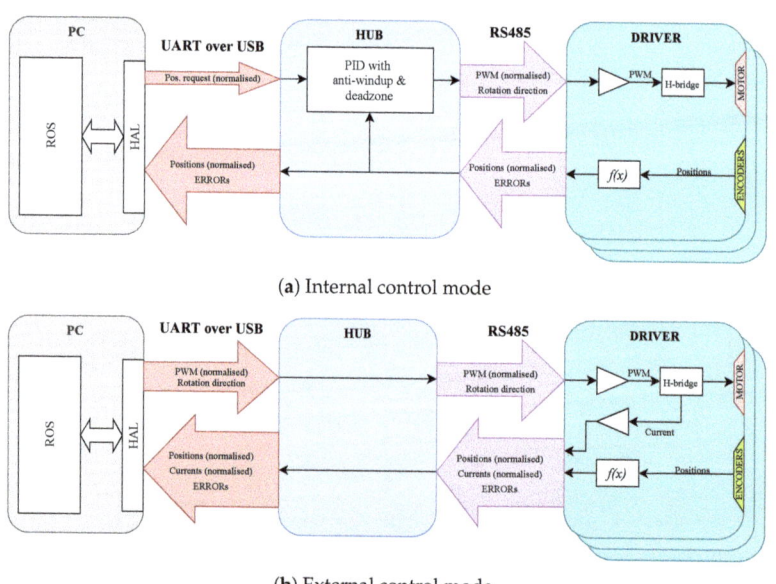

(**a**) Internal control mode

(**b**) External control mode

Figure 12. Schematics of PUT-Hand control system in selected modes.

All DRIVER modules feature a calibration procedure, which can be triggered by a high-level controller. During the procedure, the DRIVE module acts independently and moves the joint to both extreme positions to determine position sensor border values and the direction of the motor rotation. Calibration data is then stored in the EEPROM memory of the DRIVE module.

3.4. Kinematic Model of PUT-Hand

The kinematic model of the hand is a simplified version of human hand kinematics and a direct result of developed mechanical design. All digits consist of three rotational joints. In the fingers, the joints (MCP, PIP, and DIP) share a common plane. The fingers' planes are slightly spread outwards to facilitate spherical grasping.

The thumb creates a more complex kinematic chain, with its CMC joint placed at an angle, and the remaining joints (MCP and IP) sharing a common plane. CMC axis orientation was chosen to allow a widest possible range of grasp types and thumb opposition with a simple rotary joint.

Kinematic structure of the gripper together with joint angle naming is presented in Figure 13. All joint ranges are given in Table 4. The range of movements allows the fingers to fully extend and flex (touching palm with the fingertips). CMC range enables full thumb opposition, which makes precision grasping easier (thumb, index, and middle fingers meet each other).

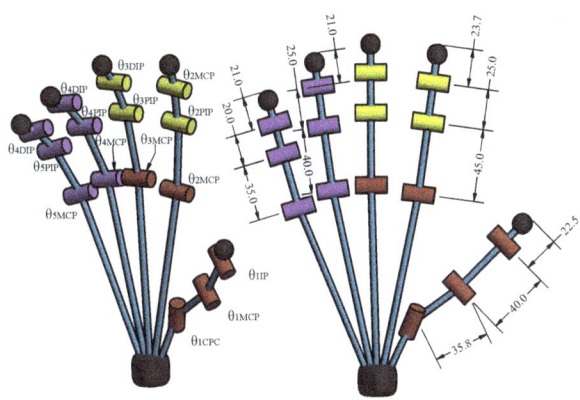

Figure 13. PUT-Hand kinematics with basic dimensions and joint naming: red—fully actuated joints, yellow—dependent joints, violet—underactuated joints.

Table 4. Joint angle ranges.

Joint	Max	Min
θ_{1CMC}	0°	135°
θ_{1MCP}	−53.8°	53.8°
θ_{1IP}	0°	90°
$\theta_{2MCP}/\theta_{3MCP}$	7°	90°
$\theta_{2PIP}/\theta_{3PIP}$	0°	96°
$\theta_{2DIP}/\theta_{3DIP}$	0°	74.2°
$\theta_{4MCP}/\theta_{5MCP}$	0°	90°
$\theta_{4PIP}/\theta_{5PIP}$	0°	98.9°
$\theta_{4DIP}/\theta_{5DIP}$	0°	99.8°

3.5. High-Level Controller

High-level control of the PUT-Hand is based on the ROS, with core driver written in C++. The driver communicates with the on-board main controller and receives information about the state of the hand and re-sends motion orders. Joint states are cyclically published in the ROS topic, so they can be accessed by multiple programs (ROS nodes) at the same time. In the experiments presented in the article, the arm and the robotic hand are controlled using the ROS environment. The motion of the whole setup is planned using the MoveIt module [69]. The MoveIt module is used to compute forward and inverse kinematic models of the hand, plan the trajectory of the fingers taking into account self-collisions, and execute the trajectory. In the simulated environment and on the real hand we use the position-based interface with the linear joint trajectory that guarantees continuity at the position level only. In the Gazebo simulation, the hand is controlled by the built-in ROS joint trajectory controller. On the real hand, the ROS driver sends the goal joint positions only. The joint trajectories result from the PID controllers in the DRIVER.

A URDF model of the PUT-Hand was also defined, containing information about kinematic model, visual shape of the links and collision model of the gripper. The URDF model is used in Gazebo simulations. The gripper can be attached to a robotic arm (in our case Universal Robots UR3). Current configuration of the PUT-Hand can be visualised using RViz. The example configuration of the hand and the corresponding visualisation in RViz are presented in Figure 14a,b, respectively.

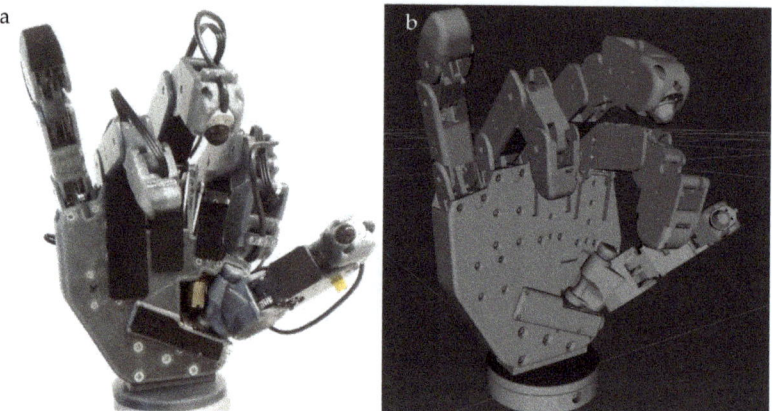

Figure 14. Sample configuration of the PUT-Hand (**a**) visualised in the RViz (**b**).

Concurrently, we have defined the most common configurations of the hand, such as initial configuration, open hand, power grasp or pinch grasp. These configurations can be quickly obtained by using predefined commands. We use them to initialise or to show motion capabilities of PUT-Hand or to recover after the error state.

4. Results

In this paragraph we first show general capabilities of the hand by performing grasp of different objects. Next, we show the use of the hand for elastic object manipulation in open-loop. Subsequently, we will present the use of force sensors attached to the hand fingertips. After, successful test of manipulation and the use of sensors we performed in-hand manipulation for elastic object identification. Finally, we demonstrate the task of inserting the plug which is often encountered in industrial setting when the cables manipulation is performed.

4.1. Grasping

To show kinematic capabilities of the hand, we presented configurations of the hand during grasping of various objects. Example grasps are shown in Figure 15. Objects with various shapes and dimensions were chosen for demonstration: pen, screwdriver, tape, saucer, cup, bottle, plastic plate, ball. We tested precise grasps (Figure 15a,c–f,h) and power grasps (Figure 15b,g,i,j). The hybrid mechanical design of PUT-Hand enables both precision manipulation with fully actuated fingers and stable power grasps with the help of underactuated digits. The underactuated digits have two main advantages. Firstly, they stabilise the position of the large or heavy objects (Figure 15d,g). Secondly, they adapt automatically to the shape of objects without the use of sophisticated tactile feedback and control systems (Figure 15j). The precision grasps with the fully actuated digits are additionally supported by the feedback from tactile sensors mounted on the fingertips of fully actuated fingers.

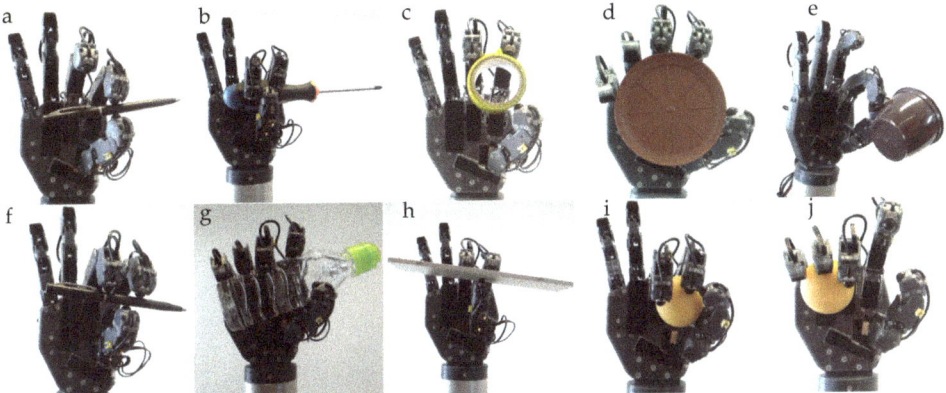

Figure 15. Sample configuration of PUT-Hand during grasping of various objects. Examples of precise grasps (**a,c–f,h**), and power grasps (**b,g,i,j**).

4.2. Elastic Object Insertion

In the first experiment, we use PUT-Hand attached to a robotic arm (UR3) to bend and force the elastic pipe into a S-shaped channel. The channel has a circular cross section, with an opening at the top narrower than its diameter. Thus, to insert the pipe into the channel the robot has to apply force until the he pipe locks in the channel. The initial configuration of the hand and the pipe are presented in Figure 16a. During the experiment, the robot uses two fully actuated fingers only (index and middle finger).

At the beginning of the experiment, the two fingers are used to force the centre of the pipe into the channel (Figure 16b). Then the procedure uses the index fingertip to bend the pipe (Figure 16c,d). The same procedure is used to bend the second side (Figure 16f–i). Finally, we use a few effective steps to bend the pipe and put it in the channel. The results are presented in Figure 16l. During the experiment, haptic feedback was not used.

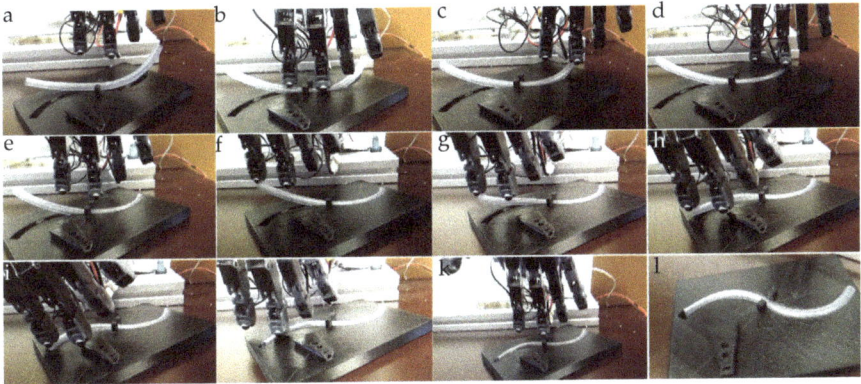

Figure 16. Inserting an elastic pipe into a channel using PUT-Hand attached to a UR3 robot: bending and pushing pipe into the channel (**a–k**), final result (**l**).

4.3. Contact Force Measurements

In the third set of experiments, we verify usability of the force sensors attached to the fingertips. As the model of the hand is defined in ROS, the direction of contact forces can be easily determined in 3D space. The position of each fingertip and corresponding contact forces can be visualised and are available at any time for the control modules. The example visualisation is available in Figure 17.

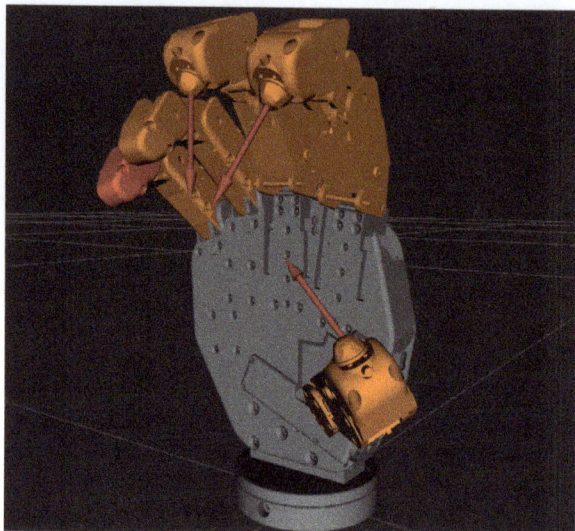

Figure 17. RViz visualisation of measured contact forces; red arrow represents the force, with arrow length corresponding to its magnitude

4.3.1. In-Hand Elastic Object Identification

In the first contact force experiment, we evaluate the possibility of stable contact event detection using reaction forces. A pinch grasp was performed using fully actuated fingers. The fingers were flexed until a specified force threshold was exceeded. Six objects with various physical properties were used: metal pipe, rubber pipe, squash ball, rubber eraser, and two types of foam. The procedure is presented in Figure 18.

An example trajectory of the thumb tip and corresponding tactile force during pinching of the metal pipe and green foam are presented in Figure 19a and Figure 19b, respectively. Tactile information from the thumb was used because its contact with the object was stable in all of the experiments, and the event of touching the object can be easily distinguished. In case of a rigid object (metal blue pipe), the contact force oscillates after the first contact (Figure 19a). In the experiment with green foam, the soft object dampens the reaction forces and they increase gradually. Obtained force trajectory show a clear difference between a rigid and a soft object and can be used for identification purposes. In both cases, the reaction force stabilises at a constant level (0.8 N for the blue pipe and 0.55 N for the green foam).

We also checked if the force sensors in the hand can identify the properties of manipulated object, such as object stiffness, during in-hand manipulation. After pinching the object, the hand changes its configuration to increase and decrease the contact forces periodically, and the results are logged. A sample relation between contact force and the displacement of the thumb tip during manipulation of the white rubber eraser is presented in Figure 20. The data shows that the displacement was lower than one millimetre, which is too small to determine the object stiffness accurately.

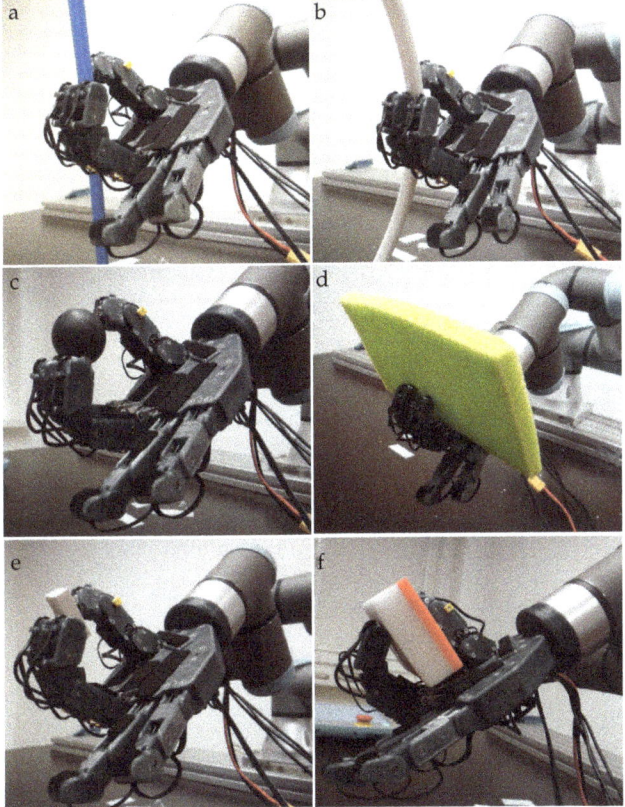

Figure 18. Objects grasped by the hand using force feedback: metal pipe (**a**), rubber pipe (**b**), squash ball (**c**), rubber eraser (**e**), and two types of foam (**d**,**f**).

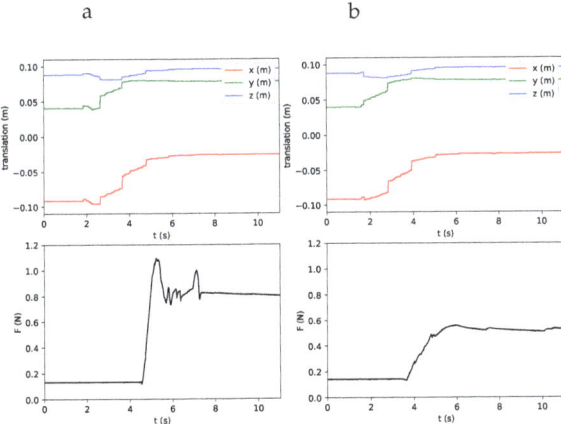

Figure 19. Trajectory of the thumb tip and corresponding tactile force trajectory during pinching the blue pipe from Figure 18a (**a**) and green foam from Figure 18d (**b**).

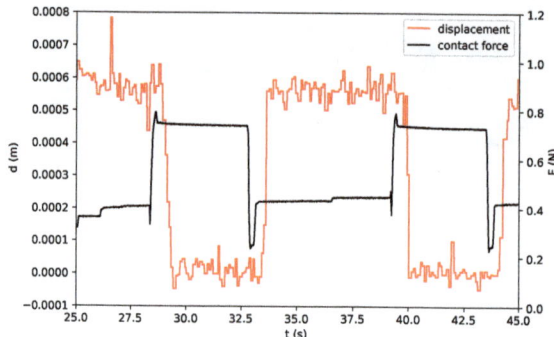

Figure 20. Contact force for the thumb and the displacement d of the thumb tip during manipulating the white rubber (Figure 18e).

4.3.2. Plug Insertion

In the second experiment, a setup consisting of PUT-Hand and UR3 arm performs a task of plug insertion. This simulates a procedure commonly used in automotive industry, widely performed by human operators. Figure 21b shows a plug and socket contraption used during the experiment. Plug contains an elastic component which bends during insertion. Socket is equipped with triangular components which centre the plug and lock it inside.

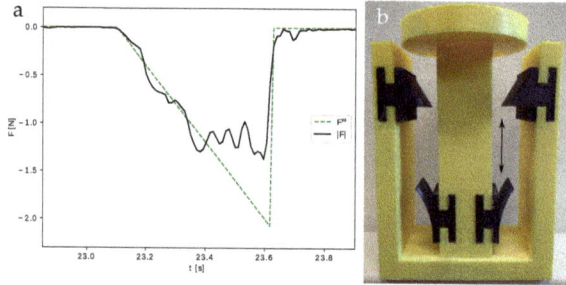

Figure 21. Forces (F) measured during plug insertion experiment and its predicate values (F^m) (**a**); photo of plug and socket pair used during the experiment (**b**).

Figure 22 shows a plug insertion procedure, in which the plug is hold with a tree-point spherical grasp using all tree fully actuated fingers. However, the thumb and the index finger generate opposite forces which hold the plug. The middle finger support the grasp only and prevents the unwanted rotation of the plug. In this experiment, during the insertion itself, main reaction forces are lateral to the tactile sensor base. While forcing two connection elements together, the elastic component of the plug bends, increasing the force required to further move the gripper.

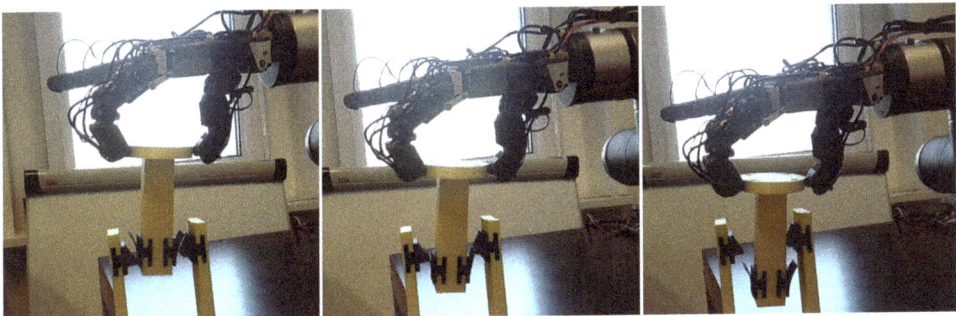

Figure 22. UR3 robot equipped with PUT-Hand inserting the plug into the socket.

To model the reaction forces in described setup we use a mechanical system presented in Figure 23. Vertical motion of the plug compresses the spring, increasing a reaction force in the direction opposite to the movement. The reaction force is proportional to the spring constant k, friction coefficient μ and the spring displacement x. The compression of the spring x depends on the motion of the plug y and the ramp angle θ:

$$\mathbf{F}^m = 2 \cdot k \cdot \mu \cdot x = 2 \cdot \frac{k \cdot \mu \cdot y}{\tan \theta}, \tag{1}$$

The measured θ angle and spring constant k is equal to $31°$ and $800 \left[\frac{N}{m}\right]$, respectively. Then, we estimated the friction coefficient μ to fit the model to the data obtained during the experiment.

Figure 21a presents reaction force $\mathbf{F}^m$ predicted using a model in Figure 23 and those measured during the experiment of plug insertion. The module of the reaction force $|\mathbf{F}|$ in Figure 21a is obtained from the forces measured at the thumb tip $\mathbf{F}^t$ and the index finger tip $\mathbf{F}^i$:

$$|\mathbf{F}| = |\mathbf{F}^t + \mathbf{F}^i|, \tag{2}$$

In the experiment we show that reaction forces which are lateral to the sensor can be measured using the given configuration of the sensors in the fingertips. The middle finger is used only to stabilise the grasp and the tactile sensor does not touch the object during the experiment. We model the static properties of the system only thus the force decreases instantly when the plug is in the socket. In practice the mass of the system and elasticities cause the gradual decrease of the reaction forces shown in Figure 21a.

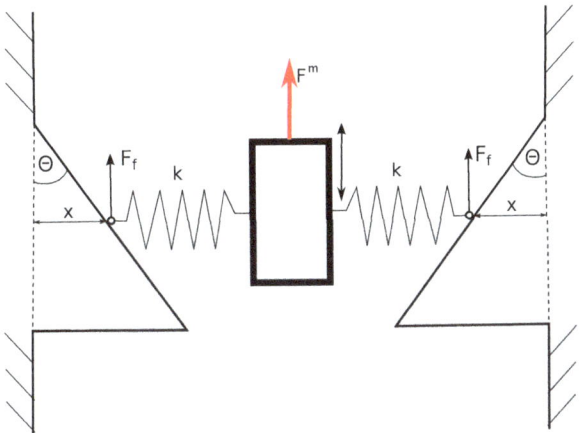

Figure 23. Mechanical model of plug insertion experimental setup.

5. Conclusions and Further Development

In this article, we propose a design of a five-finger anthropomorphic gripper intended for elastic object manipulation. All design files are published at https://github.com/puthand. Main contributions of this work are:

- open-source mechanical design of PUT-Hand, a hybrid anthropomorphic gripper design consisting of three fully actuated fingers (thumb, index, and middle) for precise manipulation, and two underactuated tendon-driven digits (ring and small) for power-grasp support;
- open-source on-board controller design and firmware;
- sensory system using optical 3-axis force sensors;
- ROS-based open-source driver for high-level control including motion planning and visualisation;
- experimental verification presenting mechanical and sensory capabilities of the proposed design.

PUT-Hand is a hybrid design, taking advantages of both fully actuated [1,5,11] and underactuated [6,24] designs. Fully actuated fingers can be used for precise grasping (Figure 15) and interaction with elastic objects (Figure 16), while underactuated fingers stabilise the grasp during dealing with heavy objects such as bottles (Figure 15g) or large objects.

The gripper is equipped with three tri-axial force sensors mounted on fingertips of fully actuated fingers, allowing for measurement of magnitude and direction of contact forces. This information can be used to measure the physical properties of the objects the robot is interacting with. Moreover, with simple modelling, we can detect the state of the environment (Figure 22).

Future work includes integration of presented setup (robot equipped with PUT-Hand gripper) with a visual perception system. Combination of visual and tactile feedback can be used to further increase the autonomy of the robot, and improve the performance of grasping, modelling and manipulation of elastic objects.

Author Contributions: Conceptualisation, T.M., J.T. and D.B.; Funding acquisition, K.W.; Investigation, T.M., J.T. and D.B.; Methodology, D.B.; Project administration, K.W.; Resources, K.W.; Software, T.M., J.T. and D.B.; Writing—original draft, T.M., J.T. and D.B.; Writing—review & editing, K.W. All authors have read and agreed to the published version of the manuscript.

Funding: This work is supported by grant No. LIDER/3/0183/L-7/15/NCBR/2016 funded by The National Centre for Research and Development (Poland)

Conflicts of Interest: The authors declare no conflict of interest.

Abbreviations

The following abbreviations are used in this manuscript:

DIP	Distant Interphalangeal joint
MCP	Metacarpophalangeal joint
IP	Interphalangeal joint
ROS	Robot Operating System
DoF	Degree of Freedom
URDF	Universal Robot Description Format

References

1. Schunk. Servo-Electric 3-Finger Gripping Hand (SDH). Available online: http://www.schunk-modular-robotics.com (accessed on 18 June 2019).
2. Zollo, L.; Roccella, S.; Guglielmelli, E.; Carrozza, M.; Dario, P. Biomechatronic Design and Control of an Anthropomorphic Artificial Hand for Prosthetic and Robotic Applications. *IEEE/ASME Trans. Mechatron.* **2007**, *12*, 418–429. [CrossRef]
3. Massa, B.; Roccella, S.; Carrozza, M.; Dario, P. Design and development of an underactuated prosthetic hand. In Proceedings of the 2002 IEEE International Conference on Robotics and Automation (Cat. No.02CH37292), Washington, DC, USA, 11–15 May 2002; pp. 3374–3379.

4. VINCENTevolution 2 Prosthetic Hand. Available online: http://vincentsystems.de/en/prosthetics/ (accessed on 18 June 2019).
5. Liu, H.; Wu, K.; Meusel, P.; Seitz, N.; Hirzinger, G.; Jin, M.; Liu, Y.; Fan, S.; Lan, T.; Chen, Z. Multisensory five-finger dexterous hand: The DLR/HIT Hand II. In Proceedings of the 2008 IEEE/RSJ International Conference on Intelligent Robots and Systems, Nice, France, 22–26 September 2008; pp. 3692–3697.
6. Aukes, D.; Heyneman, B.; Ulmen, J.; Stuart, H.; Cutkosky, M.; Kim, S.; Garcia, P.; Edsinger, A. Design and testing of a selectively compliant underactuated hand. *Int. Robot. Res.* **2014**, *33*, 721–735. [CrossRef]
7. iLimb Prosthetic Hands. Available online: http://www.touchbionics.com/products (accessed on 18 June 2019).
8. Michelangelo Prosthetic Hand. Available online: https://www.ottobockus.com/prosthetics/upper-limb-prosthetics/ (accessed on 18 June 2019).
9. Jacobsen, S.; Iversen, E.; Knutti, D.; Johnson, R.; Biggers, K. Design of the Utah/M.I.T. Dextrous Hand. In Proceedings of the 1986 IEEE International Conference on Robotics and Automation, San Francisco, CA, USA, 7–10 April 1986; Volume 3, pp. 1520–1532.
10. Martin, J.; Grossard, M. Design of a fully modular and backdrivable dexterous hand. *Int. Robot. Res.* **2014**, *33*, 783–798. [CrossRef]
11. Quigley, M.; Salisbury, C.; Ng, A.; Salisbury, J. Mechatronic design of an integrated robotic hand. *Int. Robot. Res.* **2014**, *33*, 706–720. [CrossRef]
12. Rothling, F.; Haschke, R.; Steil, J.; Ritter, H. Platform portable anthropomorphic grasping with the bielefeld 20-DOF shadow and 9-DOF TUM hand. In Proceedings of the 2007 IEEE/RSJ International Conference on Intelligent Robots and Systems, San Diego, CA, USA, 29 October–2 November 2007; pp. 2951–2956.
13. Su, Y.; Wu, Y.; Lee, K.; Du, Z.; Demiris, Y. Robust grasping for an under-actuated anthropomorphic hand under object position uncertainty. In Proceedings of the 2012 12th IEEE-RAS International Conference on Humanoid Robots (Humanoids 2012), Osaka, Japan, 29 November–1 December 2012; pp. 719–725.
14. Tomovic, R.; Boni, G. An adaptive artificial hand. *IRE Trans. Autom. Control* **1962**, *7*, 3–10. [CrossRef]
15. Treratanakulwong, T.; Kaminaga, H.; Nakamura, Y. Low-friction tendon-driven robot hand with carpal tunnel mechanism in the palm by optimal 3D allocation of pulleys. In Proceedings of the 2014 IEEE International Conference on Robotics and Automation (ICRA), Hong Kong, China, 31 May–7 June 2014; pp. 6739–6744.
16. Pons, J.; Rocon, E.; Ceres, R.; Reynaerts, D.; Saro, B.; Levin, S.; Van Moorleghem, W. The MANUS-HAND Dextrous Robotics Upper Limb Prosthesis: Mechanical and Manipulation Aspects. *Auton. Robot.* **2004**, *16*, 143–163. [CrossRef]
17. Lotti, F.; Tiezzi, P.; Vassura, G.; Biagiotti, L.; Palli, G.; Melchiorri, C. Development of UB Hand 3: Early Results. In Proceedings of the 2005 IEEE International Conference on Robotics and Automation, Barcelona, Spain, 18–22 April 2005; pp. 4488–4493.
18. Controzzi, M.; Cipriani, C.; Carrozza, C. Mechatronic Design of a Transradial Cybernetic Hand. In Proceedings of the 2008 IEEE/RSJ International Conference on Intelligent Robots and Systems, Nice, France, 22–26 September 2008; pp. 576–581.
19. Cipriani, C.; Controzzi, M.; Carrozza, M. Objectives, criteria and methods for the design of the SmartHand transradial prosthesis. *Robotica* **2010**, *28*, 919–927. [CrossRef]
20. Kamikawa, Y.; Maeno, T. Underactuated five-finger prosthetic hand inspired by grasping force distribution of humans. In Proceedings of the 2008 IEEE/RSJ International Conference on Intelligent Robots and Systems, IROS, Nice, France, 22–26 September 2008; pp. 717–722.
21. Dalley, S.; Wiste, T.; Withrow, T.; Goldfarb, M. Design of a Multifunctional Anthropomorphic Prosthetic Hand With Extrinsic Actuation. *IEEE/ASME Trans. Mechatron.* **2009**, *14*, 699–706. [CrossRef]
22. Xu, Z.; Kumar, V.; Todorov, E. A Low-cost and Modular, 20-DOF Anthropomorphic Robotic Hand: Design, Actuation and Modeling. In Proceedings of the 2013 13th IEEE-RAS International Conference on Humanoid Robots (Humanoids), Atlanta, GA, USA, 15–17 October 2013; pp. 368–375.
23. Xu, Z.; Todorov, E. Design of a highly biomimetic anthropomorphic robotic hand towards artificial limb regeneration. In Proceedings of the 2016 IEEE International Conference on Robotics and Automation (ICRA), Stockholm, Sweden, 16–21 May 2016; pp. 3485–3492.
24. Santina, C.; Grioli, G.; Catalano, M.; Brando, A.; Bicchi, A. Dexterity augmentation on a synergistic hand: The Pisa/IIT SoftHand+. In Proceedings of the 2015 IEEE-RAS 15th International Conference on Humanoid Robots (Humanoids), Seoul, Korea, 3–5 November 2015; pp. 497–503.

25. Ma, R.R.; Odhner, L.U.; Dollar, A.M. A modular, open-source 3D printed underactuated hand. In Proceedings of the 2013 IEEE International Conference on Robotics and Automation, Karlsruhe, Germany, 6–10 May 2013; pp. 2737–2743.
26. Salvietti, G.; Hussain, I.; Malvezzi, M.; Prattichizzo, D. Design of the Passive Joints of Underactuated Modular Soft Hands for Fingertip Trajectory Tracking. *IEEE Robot. Autom. Lett.* **2017**, *2*, 2008–2015. [CrossRef]
27. Mianowski, K.; Berns, K.; Hirth, J. The artificial hand with elastic fingers for humanoid robot ROMAN. In Proceedings of the 2013 18th International Conference on Methods & Models in Automation & Robotics (MMAR), Miedzyzdroje, Poland, 26–29 August 2013; pp. 448–453.
28. Dechev, N.; Cleghorn, W.; Naumann, S. Multiple Finger, passive adaptive grasp prosthetic hand. *Mech. Mach. Theory* **2001**, *36*, 1157–1173. [CrossRef]
29. Light, C.; Chappel, P. Development of a lightweight and adaptable multiple-axis hand prosthesis. *Med. Eng. Phys.* **2000**, *22*, 679–684. [CrossRef]
30. Deimel, R.; Brock, O. A Novel Type of Compliant, Underactuated Robotic Hand for Dexterous Grasping. *Proc. Robot. Sci. Syst.* **2014**, *35*, 161–185.
31. Gaiser, I.; Pylatiuk, C.; Schulz, S.; Kargov, A.; Oberle, R.; Werner, T. The FLUIDHAND III: A Multifunctional Prosthetic Hand. *JPO J. Prosthetics Orthot.* **2009**, *21*, 91–96. [CrossRef]
32. Schunk, Servo-Electric 5-Finger Gripping Hand (SVH). Available online: http://www.schunk-modular-robotics.com (accessed on 18 June 2019).
33. Losier, Y.; Clawson, A.; Wilson, A.; Scheme, E.; Englehart, K.; Kyberd, P.; Hudgins, B. An Overview of the UNB Hand System. In Proceedings of the MyoElectric Controls/Powered Prosthetics, Fredericton, NB, Canada, 14–19 August 2011; pp. 1–4.
34. You, W.S.; Lee, Y.H.; Oh, H.S.; Kang, G.; Choi, H.R. Design of a 3D-printable, robust anthropomorphic robot hand including intermetacarpal joints. *Intell. Serv. Robot.* **2019**, *12*, 1–16. [CrossRef]
35. Yang, H.; Wei, G.; Ren, L. Design and Development of a Linkage-Tendon Hybrid Driven Anthropomorphic Robotic Hand. In *Intelligent Robotics and Applications*; Yu, H., Liu, J., Liu, L., Ju, Z., Liu, Y., Zhou, D., Eds.; Springer International Publishing: Cham, Switzerland, 2019; pp. 117–128.
36. Mizushima, K.; Oku, T.; Suzuki, Y.; Tsuji, T.; Watanabe, T. Multi-fingered robotic hand based on hybrid mechanism of tendon-driven and jamming transition. In Proceedings of the 2018 IEEE International Conference on Soft Robotics (RoboSoft), Livorno, Italy, 24–28 April 2018; pp. 376–381.
37. Yazici, M.V.; Kahveci, A.; Kiziltaş, F.S.; Mülayim, N.; Gezgin, E. Design and Development of a Surgical Robotic Hand with Hybrid Structure. In Proceedings of the 2018 Medical Technologies National Congress (TIPTEKNO), Magusa, Cyprus, 8–10 November 2018; pp. 1–4.
38. Mahanta, G.B.; Rout, A.; Deepak, B.B.V.L.; Biswal, B.B.; Gunji, B.M. Preliminary Design and Fabrication of Bio-Inspired Low-Cost Hybrid Soft-Rigid Robotic Hand for Grasping Delicate Objects. In Proceedings of the 2019 9th Annual Information Technology, Electromechanical Engineering and Microelectronics Conference (IEMECON), Jaipur, India, 13–15 March 2019; pp. 17–23.
39. Jeong, H.; Cheong, J. Design of hybrid type robotic hand : The KU hybrid HAND. In Proceedings of the 2011 11th International Conference on Control, Automation and Systems, Gyeonggi-do, Korea, 26–29 October 2011; pp. 1113–1116.
40. Jeong, H.; Cheong, J. Design and analysis of KU hybrid hand—Type II. In Proceedings of the 2013 10th International Conference on Ubiquitous Robots and Ambient Intelligence (URAI), Jeju, Korea, 30 October–2 November 2013; pp. 580–583.
41. Cerruti, G.; Chablat, D.; Gouaillier, D.; Sakka, S. ALPHA: A hybrid self-adaptable hand for a social humanoid robot. In Proceedings of the 2016 IEEE/RSJ International Conference on Intelligent Robots and Systems (IROS), Daejeon, Korea, 9–14 October 2016; pp. 900–906.
42. Townsend, W. The BarrettHand grasper—programmably flexible part handling and assembly. *Ind. Robot.* **2000**, *27*, 181–188. [CrossRef]
43. Aukes, D.; Cutkosky, M. Simulation-based tools for evaluating underactuated hand designs. In Proceedings of the 2013 IEEE International Conference on Robotics and Automation, Karlsruhe, Germany, 6–10 May 2013; pp. 2067–2073.

44. Grioli, G.; Catalano, M.; Silvestro, E.; Tono, S.; Bicchi, A. Adaptive synergies: An approach to the design of under-actuated robotic hands. In Proceedings of the 2008 IEEE/RSJ International Conference on Intelligent Robots and Systems, IROS, Vilamoura, Portugal, 7–12 October 2012; pp. 1251–1256.
45. Xu, K.; Liu, H. Continuum Differential Mechanisms and Their Applications in Gripper Designs. *IEEE Trans. Robot.* **2016**, *32*, 754–762. [CrossRef]
46. Tomczyński, J.; Mańkowski, T.; Walas, K.; Kaczmarek, P. CIE-Hand towards Prosthetic Limb. In *Progress in Automation, Robotics and Measuring Techniques*; Szewczyk, R., Zieliński, C., Kaliczyńska, M., Eds.; Springer International Publishing: Cham, Switzerland, 2015; pp. 275–284.
47. Kopicki, M.; Detry, R.; Adjigble, M.; Stolkin, R.; Leonardis, A.; Wyatt, J.L. One-shot learning and generation of dexterous grasps for novel objects. *Int. J. Robot. Res.* **2015**, *35*, 959–976. [CrossRef]
48. Kopicki, M.; Belter, D.; Wyatt, J.L. Learning better generative models for dexterous, single-view grasping of novel objects. *Int. J. Robot. Res.* **2019**. [CrossRef]
49. Arruda, E.; Wyatt, J.; Kopicki, M. Active vision for dexterous grasping of novel objects. In Proceedings of the 2016 IEEE/RSJ International Conference on Intelligent Robots and Systems (IROS), Daejeon, Korea, 9–14 October 2016; pp. 2881–2888.
50. Grebenstein, G.; Suchi, M.; Kampel, M.; Vincze, M. An Empirical Evaluation of Ten Depth Cameras: Bias, Precision, Lateral Noise, Different Lighting Conditions and Materials, and Multiple Sensor Setups in Indoor Environments. *IEEE Robot. Autom. Mag.* **2019**, *26*, 67–77.
51. Tian, S.; Ebert, F.; Jayaraman, D.; Mudigonda, M.; Finn, C.; Calandra, R.; Levine, S. Manipulation by Feel: Touch-Based Control with Deep Predictive Models. In Proceedings of the IEEE International Conference on Robotics and Automation, Montreal, QC, Canada, 20–24 May 2019; pp. 818–824.
52. Hsiao, K.; Chitta, S.; Ciocarlie, M.T.; Jones, E.G. Contact-reactive grasping of objects with partial shape information. In Proceedings of the 2010 IEEE/RSJ International Conference on Intelligent Robots and Systems, Taipei, Taiwan, 18–22 October 2010; pp. 1228–1235.
53. Lin, J.; Calandra, R.; Levine, S. Learning to Identify Object Instances by Touch: Tactile Recognition via Multimodal Matching. In Proceedings of the IEEE International Conference on Robotics and Automation, Montreal, QC, Canada, 20–24 May 2019; pp. 3644–3650.
54. Hyttinen, E.; Kragic, D.; Detry, R. Learning the tactile signatures of prototypical object parts for robust part-based grasping of novel objects. In Proceedings of the IEEE International Conference on Robotics and Automation, Seattle, WA, USA, 26–30 May 2015; pp. 4927–4932.
55. Tomo, T.P.; Schmitz, A.; Wong, W.K.; Kristanto, H.; Somlor, S.; Hwang, J.; Jamone, L.; Sugano, S. Covering a robot fingertip with uskin: A soft electronic skin with distributed 3-axis force sensitive elements for robot hands. *IEEE Robot. Autom. Lett.* **2018**, *2*, 124–131. [CrossRef]
56. Funabashi, S.; Yan, G.; Geier, A.; Schmitz, A.; Ogata, T.; Sugano, S. Morphology-Specific Convolutional Neural Networks for Tactile Object Recognition with a Multi-Fingered Hand. In Proceedings of the IEEE International Conference on Robotics and Automation, Montreal, QC, Canada, 20–24 May 2019; pp. 57–63.
57. Blanes, C.; Mellado, M.; Beltrán, P. Tactile sensing with accelerometers in prehensile grippers for robots. *Mechatronics* **2016**, *33*, 1–12. [CrossRef]
58. Seminara, L.; Pinna, L.; Ibrahim, A.; Noli, L.; Caviglia, S.; Gastaldo, P.; Valle, M. Towards integrating intelligence in electronic skin. *Mechatronics* **2016**, *34*, 84–94. [CrossRef]
59. Dargahi, J.; Sedaghati, R.; Singh, H.; Najarian, S. Modeling and testing of an endoscopic piezoelectric-based tactile sensor. *Mechatronics* **2007**, *17*, 462–467. [CrossRef]
60. Pan, Z.; Zhu, Z. Flexible full-body tactile sensor of low cost and minimal output connections for service robot. *Mechatronics* **2005**, *32*, 485–491. [CrossRef]
61. Hristu, D.; Ferrier, N.; Brockett, R.W. The performance of a deformable-membrane tactile sensor: Basic results on geometrically-defined tasks. In Proceedings of the IEEE International Conference on Robotics and Automation, San Francisco, CA, USA, 24–28 April 2000; pp. 508–513.
62. Hashizume, J.; Tae Myung, H.; Suresh, S.A.; Cutkosky, M.R. Capacitive Sensing for a Gripper with Gecko-Inspired Adhesive Film. *IEEE Robot. Autom. Lett.* **2019**, *4*, 677–683. [CrossRef]
63. Fearing, R. Tactile sensing mechanisms. *Int. J. Robot. Res.* **1990**, *9*, 3–23. [CrossRef]
64. Luo, S.; Bimbo, J.; Dahiya, R.; Liua, H. Robotic tactile perception of object properties: A review. *Mechatronics* **2017**, *48*, 54–67. [CrossRef]

65. Wettels, N.; Fishel, J.; Loeb, G. Multimodal Tactile Sensor. In *The Human Hand as an Inspiration for Robot Hand Development, Springer Tracts in Advanced Robotics*; Ramachandran Balasubramanian, S.V., Ed.; Springer: Cham, Switzerland, 2016; pp. 405–429.
66. Feix, T.; Romero, J.; Schmiedmayer, H.; Dollar, A.M.; Kragic, D. The GRASP Taxonomy of Human Grasp Types. *IEEE Trans. Hum. Mach. Syst.* **2016**, *46*, 66–77. [CrossRef]
67. Johnson, P.W.; Blackstone, J.M. Children and gender-differences in Exposure and How Anthropometric Differences Can be Incorporated into the Design of Computer Input Devices. *Scand. J. Work. Environ. Health* **2007**, *26*, 26–32.
68. Scalera, L.; Palomba, I.; Wehrle, E.; Gasparetto, A.; Vidoni, R. Natural Motion for Energy Saving in Robotic and Mechatronic Systems. *Appl. Sci.* **2019**, *9*, 3516. [CrossRef]
69. Coleman, D.; Sucan, I.; Chitta, S.; Correll, N. Reducing the Barrier to Entry of Complex Robotic Software: A MoveIt! Case-Study. *J. Softw. Eng. Robot.* **2014**, *5*, 3–16.

© 2020 by the authors. Licensee MDPI, Basel, Switzerland. This article is an open access article distributed under the terms and conditions of the Creative Commons Attribution (CC BY) license (http://creativecommons.org/licenses/by/4.0/).

Article

On Robustness of Multi-Modal Fusion—Robotics Perspective

Michal Bednarek * **, Piotr Kicki and Krzysztof Walas**

Institute of Robotics and Machine Intelligence, Poznan University of Technology, 60-965 Poznan, Poland; piotr.kicki@put.poznan.pl (P.K.); krzysztof.walas@put.poznan.pl (K.W.)
* Correspondence: michal.bednarek@put.poznan.pl

Received: 16 June 2020; Accepted: 10 July 2020; Published: 16 July 2020

Abstract: The efficient multi-modal fusion of data streams from different sensors is a crucial ability that a robotic perception system should exhibit to ensure robustness against disturbances. However, as the volume and dimensionality of sensory-feedback increase it might be difficult to manually design a multimodal-data fusion system that can handle heterogeneous data. Nowadays, multi-modal machine learning is an emerging field with research focused mainly on analyzing vision and audio information. Although, from the robotics perspective, haptic sensations experienced from interaction with an environment are essential to successfully execute useful tasks. In our work, we compared four learning-based multi-modal fusion methods on three publicly available datasets containing haptic signals, images, and robots' poses. During tests, we considered three tasks involving such data, namely grasp outcome classification, texture recognition, and—most challenging—multi-label classification of haptic adjectives based on haptic and visual data. Conducted experiments were focused not only on the verification of the performance of each method but mainly on their robustness against data degradation. We focused on this aspect of multi-modal fusion, as it was rarely considered in the research papers, and such degradation of sensory feedback might occur during robot interaction with its environment. Additionally, we verified the usefulness of data augmentation to increase the robustness of the aforementioned data fusion methods.

Keywords: multi-modal fusion; machine learning; robotics

1. Introduction

A dynamic fusion of multi-modal information is a key ability that humans utilize for a wide variety of tasks that demand an understanding of the physical properties of objects. For example, we fuse visual and haptic data to manipulate dexterously, recognize unknown objects, or localize them in the scene. However, when we want go to the bathroom at night and we want to open the water tap, somehow we know that an image under these conditions is not reliable and we should focus more on our other senses. By touching the tap we can (to some extent) supersede the vision in the localization task and perform the same task without it. The interaction between information from different senses was observed and tested experimentally in [1], where a multi-sensory illusion was called the McGurk effect. In the robotics field, a typical approach to the multi-modal data fusion is through various probabilistic models that are based mainly on the Bayesian inference. However, due to large volumes of available multi-modal and multi-relational datasets, that kind of analysis can be hindered. To overcome that problem machine learning approaches were proposed, as they can handle large and multidimensional data. In recent years there was a lot of research in the area of efficient fusion of data using machine learning, especially neural networks [2]. Nevertheless, researchers focused on the improvements in the accuracy of their models and paid almost no attention to their robustness to non-nominal conditions, which are ubiquitous in the robotics applications.

To bridge this gap, in our work we compared the performance of four popular multi-modal fusion methods utilizing artificial neural networks (ANN) on three datasets. Moreover, we extensively tested them in various data degradation scenarios, which simulated the typical issues like noisy data or sensor failure. Furthermore, we verified the influence of data augmentation on a fusion methods robustness. The general design of our experiments was depicted in Figure 1.

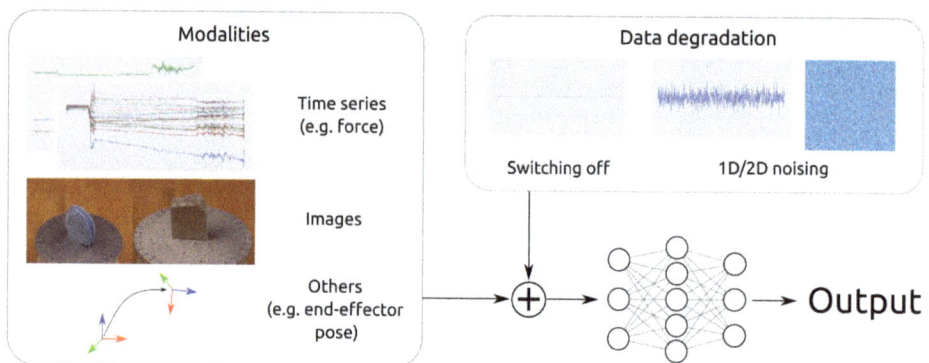

Figure 1. Experimental setup. We tested four different methods of multi-modal fusion and verified their robustness on a variety of data degradation scenarios common in robotics.

The development of approaches for fairly complicated robotic tasks, like, e.g., dexterous manipulation, nowadays very often depends more on advances in sensory systems, such as skin-like sensors [3–5] than fusion algorithms themselves. The use of fusion in dexterous manipulation [6,7] is emerging field. The more focus on development of multi-modal systems in the field of robotics is especially visible in areas like image segmentation [8–10], 3D reconstruction [11] and a tactile understanding [12].

In our work we want to stress the fact that in robotics, a multi-modal fusion is particularly important due to a possibility to improve the robustness of predictions affected by noise or failure of a sensor. Typically, it can be achieved by finding pieces of information among different modalities that exhibit interchangeability and complementarity. However, in most applications researchers focuses mostly on the improving the accuracies of their systems, by exploiting the complementarity. There is almost no consideration on interchangeability of the data sources, in the context of the robustness to, typical for real-life scenarios, noises and sensor faults.

A multi-modal machine learning is a scientific field of growing interest that brings many challenges. In [2] the authors have listed open questions to which answers should be found to advance the state of the art—how to represent the data (representation), how to map a knowledge from one modality to others (translation), how to find dependencies between heterogeneous modalities (alignment), how to join the multi-modal data stream together (fusion) and how to successfully transfer a knowledge from training a model on one modality to other (co-learning). We are aware that there are many more data fusion techniques like Kalman Filter, Bayesian Inference or Early Fusion. Unfortunately, each of those solutions is somewhat limited to low-dimensional or homogeneous data, thus we decided to exclude them from our comparison and focused on the most flexible approaches, which are model agnostic and can operate on any type of data. In our work, we evaluated a performance of multiple data fusion methods based on neural networks in robotics oriented tasks. Our contributions are:

1. Experimental evaluation of the performance of four machine-learning fusion methods—Late Fusion (Late), Mixture of Experts (MoE) [8], Intermediate Fusion (Mid) and the most recent Low-rank Multi-modal Fusion (LMF) [13]. We tested their capabilities on three multi-modal datasets in single and multi-label classification tasks.

2. Validation of the robustness of selected fusion methods to various data degradation scenarios that may occur when the robot interacts with the environment.
3. Evaluation of the influence of a data augmentation on a fusion methods performance, when a data degradation occurred during tests.

The remainder of the paper is organized as follows. First, we will provide a comprehensive review of the related work in the field of multi-modal data fusion. Then, we will present tested fusion methods and datasets used in our experiments. Next, we will move on to the results section followed by the discussion. Finally, concluding remarks will be given.

2. Related Work

The section contains a broad review of research conducted on multi-modal fusion approaches and their evaluation for data degradation, which might occur in real-world scenarios.

2.1. Data Fusion Approaches

In [2] the authors divided the multi-modal fusion techniques into two classes: model agnostic and model-based. In our paper we limited ourselves to model agnostic fusion methods as they are more general and widely spread in the robotics community. There exist three main types of model agnostic data fusion methods—early (data-level), intermediate (feature-level), and late (decision-level) [14], however it is possible to combine at least two of them into a hybrid fusion [2]. Systematic division of different sensor fusion methods is given in [15].

In early fusion, data from different modalities are typically concatenated at the early stages of data processing. It gives the machine learning model a possibility to capture even low-level interactions between modalities and process them jointly. However, that approach is limited to the cases when it is possible to concatenate the data, which is sometimes cumbersome. For example, it is not clear how to combine heterogeneous data such as 2D images with a 1D time series. For that reason, we do not compare early fusion in this paper, as it is not applicable to all data types. Typical examples of early fusion can be found in the area of semantic scene understanding using multi-spectral images [16], where visual streams from RGB, depth, and near-infrared channels are combined to produce predictions. Another example of an early fusion approach was [17]. The authors proposed to fuse RGB images with optical flow maps for gesture recognition. Other example is the use of different depth sensing modalities, with different properties to obtain denser depth image [18].

A feature-level fusion is a very popular technique in machine learning models as it merges data representations at higher levels of abstraction. That in turn allows for combining even heterogeneous data from very different sources and lets the machine learning model to process joint representation in order to produce an output. That type of fusion is widely spread in robotics community in areas such as object recognition [19] and scene recognition [20] tasks. Multi-modal fusion applied to robot motion planning was presented in [21] and contact-rich manipulation tasks in [7]. A different approach to a feature fusion is presented in [22], where instead of features concatenation, only some randomly chosen parts of feature vectors from different modalities are merged. Another intermediate fusion approach exploits Tensor Fusion Networks [23,24], which are extensively used for example in the multi-modal sentiment analysis. However, they, to the best of the authors' knowledge, are not used in the robotics applications. The main issue of these approaches is their low computational efficiency, which is addressed in the paper about Low-rank multi-modal Fusion [13], which exploits the tensor decomposition to reduce the number of model's parameters.

Similar to feature-level fusion approaches, late ones can work with any data types. However, they do not merge the data but only outputs of the models, which process different modalities separately. Prominent work on late fusion approaches is described in [25]. The typical scheme of a late fusion was presented in [26–28]. The authors of [27,28] proposed the late fusion approach to process RGB-D data in the tasks of object detection and discovery respectively. In [26] images and point clouds were used to perform semantic segmentation of the urban environment for autonomous vehicles. In [29] the authors proposed a late fusion model, which took into account the impact of a data degradation on the model's decision and used a noisy-or operation to combine these decisions. The use of three different modalities for terrain classification fused using late fusion approach is presented in [30].

If the proposed solution merges information on two or more levels, we talk about hybrid fusion. Examples of that approach are presented in [8,9] where outputs of models are fused with the use of weights determined by the gating network, which uses an intermediate representation of all modalities. That approach, called the Mixture of Experts is able to decide, which modality should have a stronger impact on the final outcome based on the features extracted from all modalities.

2.2. Fusion Robustness

Fusion robustness is a rarely considered topic in the multi-modal fusion literature, however, it seems to be an important issue in real-world applications, especially in the robotics field. Only several papers [22,29,31–34] took into account the non-nominal conditions of the multi-modal fusion and provided some analysis of fusion robustness to data degradation. Such degradation could occur due to sensor noise, its failure, or unexpected weather conditions. Moreover, the approach proposed in [8] is potentially able to take into account data degradation and express the belief in terms of the weighting of the model's decisions. However, it was not considered by the authors. Even though there are works on the robustness of multi-modal fusion in the robotic context, one can observe a lack of comprehensive joint comparison of main fusion paradigms, which were used by individual authors in the presence of sensor noises and/or failures.

3. Experiment Design

In the following section, we provided a detailed description of multi-modal datasets used in our experiments including the procedures of data preparation, cross-validation and splitting into train/test subsets. Moreover, we presented compared fusion methods with a discussion of their architectures.

3.1. Data Preparation

In our work, we measured the performance and robustness against data degradation of different fusion methods using three datasets containing multi-sensory data. First of all, from each dataset a test subset was separated and it remained unchanged throughout all of our experiments. The important note is that the test set was not involved in any training procedures described further and served only as a reference point for comparisons between methods. To ensure a fair comparison, for each dataset, we ensured that the distribution of the classes both for train and test set is similar. The same principle was maintained for the cross-validation. Each class was evenly distributed between consecutive folds using iterative stratification method [35,36]. Thus the case that some class was under or over-represented in some part of data was eliminated.

In our work, one turn of cross-validation proceeded as follows. We split the dataset into k chunks called folds. Folds numbered from 0 to $k-1$ were used for training an ANN. After that, the k-th fold was used for validation. That procedure was repeated 5 times, each time different fold was chosen as the validation set (5-fold cross validation). Moreover, in all our experiments, input data was standardized by subtracting the mean taken for all samples from a corresponding modality and divided by its standard deviation.

3.2. Multi-Modal Datasets

BioTac Grasp Stability Dataset (BiGS): The grasp-stability dataset [37] contained signals recorded during 2000 trials of grasping three types of objects—a ball, box, and a cylinder. Time series of gripper's poses and 3-axis forces gathered while shaking an object with a closed gripper. Each trial was annotated with a label *success* or *fail* that corresponds to the outcome of a trial. To gather tactile signals there were used three bio-inspired BioTac [5] tactile sensors mounted on fingers of a gripper and a force-torque sensor mounted in the wrist. In our experiments, to verify whether a signal represents a successful grasp or a failure, we used gripper's positions, orientations, and force readings from the force-torque sensor. We re-sampled each signal to fit common length using the Fourier method. Finally, each input element consisted of three-time series with a length equal to 1053. Positions are expressed as 3-element vectors, orientation is presented as a 4-element quaternion, and force reading is composed of 3 values corresponding to X, Y, and Z axes. In total, the training dataset used in the cross-validation was composed of 3197 samples for each modality, while the test set of 801.

The Penn Haptic Texture Toolkit (HaTT): In the toolkit [38] there are 100 different textures photographed and presented as RGB images. Each texture has associated normal force, acceleration, and position recordings gathered during unconstrained motions of an impedance-type haptic device SensAble Phantom Omni [39]. To save time needed for training, in our experiments we used data from 10 classes only, the names of which were presented in Table 1.

Table 1. Textures from the Penn Haptix Toolkit chosen for experiments.

ABS Plastic	Aluminium Foil	Aluminium Square	Artificial Grass	Athletic Shirt
Binder	Blanket	Book	Brick 1	Brick 2

Signals in the HaTT dataset were gathered using a haptic device's tool-tip while moving on different surfaces for 10 s. In our experiments, we used a normal force, acceleration and velocity as input modalities. However, it is important to mention that the authors of the dataset used a method called DFT321 [40] to combine 3-axes signals of acceleration and velocity into single axes, thus in our experiments, we used the 1-dimensional representations of these quantities. The authors motivated that dimensionality reduction by the fact that humans do not perceive the direction of high-frequency vibrations, which was described in [41]. We did not use available RGB images, because each class had only one associated image, thus there were far too few of them to train an ANN. Every time series was cut into vectors with a length of 200, which resulted in a total number of 8000 samples included in the training set and another 2000 in the test set. Again, we made this split maintaining the equal balance between classes.

The Penn Haptic Adjective Corpus 2 (PHAC-2): The last dataset used in our experiments considers the problem of multi-label classification of haptic adjectives using data created by the authors of [42] and dataset was further refined by authors of [12]. The dataset consisted of 53 objects photographed from 8 different directions. Each photo had corresponding haptic signals from the squeezing of an object gathered from two BioTac sensors. Moreover, every object was described with several haptic adjectives used as labels. In the dataset there were 24 haptic adjectives, in Figure 2 we presented their histogram.

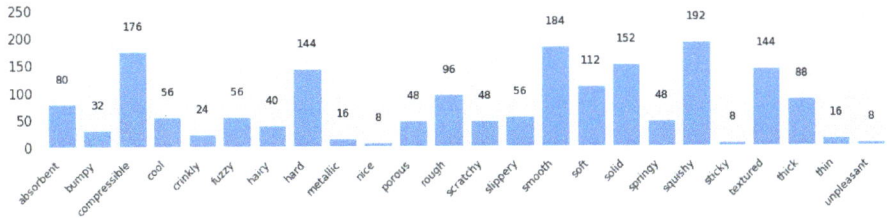

Figure 2. Occurrences of each adjective in the Penn Haptic Adjective Corpus 2 (PHAC-2) dataset.

To perform the experiments on the balanced train and test sets, we used the iterative stratification to ensure that there is no significant over or under-representation of any adjective in the train/test subsets. It is important, because such imbalance can lead to a significant drop in the prediction performance, hence giving misleading results. A single training sample was composed of an RBG image with the spatial resolution of 224 × 224 together with two raw signals from 19-electrode arrays from both of BioTac sensors. As the length of time series was different, we re-sampled them to fit a fixed number of 67 values. In total, we had 265 samples in the training dataset and 159 in the test set.

3.3. Fusion Methods

In related work we discussed several model agnostic data fusion techniques, but for the experimental analysis we selected four of them (see Figure 3). Taking into account the simplicity and popularity we decided to include in our benchmark basic versions of late and intermediate fusion. The third fusion candidate was the Mixture of Experts [8] fusion model, which was similar to late fusion, but it was able to decide on the modality importance based on their latent representations. Moreover, we also included a method that was previously not used in the robotics community but achieved some very promising results in other areas such as sentiment analysis—the Low-Rank multi-modal Fusion (LMF). To obtain fair comparison, in our implementation of aforementioned fusion techniques we first transformed input modalities to the 10-dimensional latent space. In this way we obtained N latent vectors $L_1, L_2, \ldots, L_N$, which were provided to corresponding ANNs. Approaches to data fusion, which we decided to examine in this paper, are presented schematically in Figure 3 and described in detail below.

Late Fusion: The main idea standing behind that method was to process each modality separately and merge predictions at the very end of the process assuming that they were of the same importance. The merging is performed on the decision level as it is described in [15]. Referring to Dasarathy classification this approach is described as Decision In-Decision Out (DEI-DEO). In our experiments we processed each of latent vectors separately using ANNs (represented in Figure 3 as arrows) to obtain predictions for each modality $p_1, p_2, \ldots, p_N$ in the form of logits. Next, those logits were summed up and transformed into class probabilities using a softmax function.

Mixture of Experts (MoE): The approach presented in [8] is built upon the Late Fusion method, however, it could decide on the modalities importance through the gating network. That decision was encoded in the weights vector w, such that $\sum_{i=1}^{N} w_i = 1$. In contrast to the Late Fusion, before a summation of predictions from all modalities they were multiplied by corresponding weights. Thus, the value of a vector w was determined by a relatively small fully connected neural network, which used all latent vectors and produced final predictions. That architecture potentially allowed ANN to learn how to react to the degradation of some modalities, by assigning lower weights to the degraded modalities. On the other hand, if the data degradation did not occur during a training phase, there was a possibility that the MoE would put too much emphasis on the modality affected by some noise during testing, which might result in false predictions.

Intermediate Fusion (Mid): The fusion of information carried by individual modalities was made by concatenating their representations in the latent space. Next, a common representation was processed further to obtain a joint prediction. The merging is performed on the feature level as it is described in [15]. Referring to Dasarathy classification this approach is described as Feature In-Feature Out (FEI-FEO). In our experiments joint predictions were produced by ANN. That approach allowed a fusion model to take into account data from all modalities in the latent space and process them freely. The Mid method would also be able to gain some robustness to the data degradation during the training, as it could learn to reduce the impact of degraded modalities. However, in contrast to MoE, its robustness and decisions were not so clearly interpretable.

Low-rank Modality Fusion (LMF): In our work, we used also a method very different than others. It was the tensor-based approach for multi-modal fusion, which was focused on revealing the interactions between features extracted from different modalities. Generally, the core idea of tensor approaches is a creation of some high-dimensional tensor representation by taking the outer products over the set of uni-modal latent representations. That representation is then linearly mapped to some low-dimensional space using learned weights and biases. Typically, such approaches suffer from computational inefficiency as that tensor weights and the number of multiplications scales exponentially with the number of modalities. However, the approach proposed in [13], did not multiply high-dimensional weight tensor with the tensor representation of the data directly. Instead, the authors proposed to firstly decompose tensor weights into N sets of modality-specific factors similar to the representation of the input decomposes into low-dimensional feature vectors. Such decomposition reduced the number of computations, as it let to directly map from feature space to predictions without explicitly creating any high-dimensional tensors.

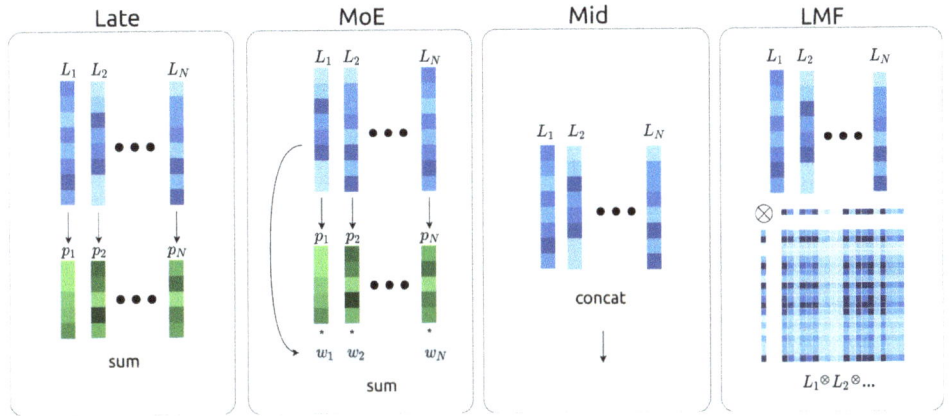

Figure 3. Multi-modal fusion architectures used in experiments. From left: Late Fusion (Late), Mixture of Experts (MoE), Intermediate Fusion (Mid), Low-rank Modality Fusion (LMF). Arrows represent the transformations realized with the use of neural networks, while L_i, p_i, w_i denotes the latent vector, predictions and trainable weight associated with i-th modality.

3.4. Neural Network Architectures

To achieve a fair comparison between fusion methods, we designed neural network architectures in a way that ensured a similar number of the trainable parameters for each type of fusion. In our experiments, we operated on two different kinds of signals—time series and images. For time-series processing we used a few 1D convolutional layers (Conv1D), followed by the Long-Short Term Memory (LSTM) units and fully connected (FC) layers, whereas for images we used 2D convolutional layers (Conv2D) with a few FC layers on top of them.

For both BiGS and HaTT datasets, we used similar architectures for determining the latent vectors. They were composed of 3 Conv1D layers with 64 filters of size 5×5 with stride equal to 2, followed by LSTM layer with 32 units, and 2 FC layers with 128 and 10 neurons respectively. However, in the case of the BiGS dataset, for Mid, Late, and MoE the number of units in the last FC layer was changed to 2, as we performed binary classification. Moreover, for both BiGS and HaTT we reduced the number of filters in all convolutional layers for MoE, as it used an additional network to produce the weights w_i. This network for all datasets had the same architecture, namely, 3 FC layers with 128, 64, and N units, where N is the number of modalities. A similar network was used in the Mid fusion to process the concatenated latent vectors into the predictions, however, in the last layer, the number of units was equal to the number of classes.

In the case of PHAC-2 dataset, we had to process both time series as well as images. Neural networks for time series had similar architecture as for BiGS and HaTT datasets, however with an increased number of neurons in last FC layer—24 and reduced number of Conv1D layers equal 2 for all methods except the LMF, which stayed with 3 Conv1D layers and 10 neurons in the last FC layer. For image processing we used 2 Conv2D layers with 64 filters of size 5×5 with stride equal to 2, followed by 2 FC layers with 128 and 24 neurons, except the LMF, which had 10 neurons in the last layer.

Additional details about the implementation of the fusion methods, as well as the architectures of the neural network used in the experiments, can be found in the code repository (https://bitbucket.org/m_bed/sense-switch/).

4. Results

In the following section, we presented results form the performance evaluation of four multi-modal fusion methods (Late, MoE, Mid, LMF) on three different datasets (BiGS, HaTT, PHAC-2). We firstly did 5-fold cross-validation (k-folds I-V) and presented results in Section 4.1. At that stage, we not only did the cross-validation but also chose the best performing models for further experiments. In tables, the chosen ones were marked with a blue color. After that, we measured the influence of multiple data degradation scenarios on the performance of each method and reported results in Section 4.2. Finally, the Section 4.3 contains the outcome of experiments conducted towards data augmentation and its influence on robustness on data degradation of each fusion method. It is important to notice that all reported results were verified on separate test subsets, which remains unchanged among fusion methods.

4.1. Comparison of Fusion Method

In the first stage of experiments, we compared the performance of fusion methods on the BiGS dataset in the grasp outcome classification—a success or a failure. Input modalities were time series of gripper positions, orientations, and 3-axis forces from a wrist-mounted force-torque sensor. The final results were presented in Table 2. We reported the mean accuracy [%] with its standard deviation among the consecutive folds. The best performing models of Late (I-fold), MoE (III), Mid (II), and LMF (I) were chosen for the stage of experiments that includes the assessment of their robustness against data degradation and influence of input data augmentation. The fact that the average results in subsequent folds were very similar means that differences in data distributions across folds were negligible.

Table 2. The comparison of four fusion methods performed on the BioTac Grasp Stability Dataset (BiGS) dataset.

	I	II	III	IV	V	Mean
Late	88.9	88.1	87.9	88.5	88.1	88.3 ± 0.4
MoE	89.0	88.3	89.1	87.6	88.4	88.5 ± 0.6
Mid	88.1	89.9	89.0	87.8	88.6	88.7 ± 0.8
LMF	89.6	88.4	88.0	88.4	88.9	88.7 ± 0.6

Cross-validation on the HaTT dataset was another step in our experiments. Results in the form the classification accuracy [%] were reported in Table 3. As input modalities, we used again time series—a squashed 1-dimensional representation of acceleration and velocity, together with a normal force acting on a haptic device's tool-tip. For the next experiments, we chose the II-fold models for the Late, MoE, LMF methods, and III-fold model for the Mid fusion approach.

In the task of multi-label classification of haptic adjectives, we used the PHAC-2 dataset. Similarly, as in the [12], we chose as a performance metric the Area Under a Curve (AUC) that measures the area under the Receiver Operating Characteristic (ROC) curve. That metric is widely spread in the multi-label classification field of machine learning. It measures how good the predictive model can distinguish between classes (in our case—haptic adjectives) taking into account a correspondence between a sensitivity/specificity ratio and multiple values of a decision threshold. In the AUC-ROC metric, a value of 1.0 refers to an excellent classification ability, 0 means that the model is always wrong, while 0.5 means that model has no discrimination capacity. In Table 4 we reported AUC-ROC metrics achieved by evaluated fusion methods. For further stages of experiments, we chose V-fold models of the Late and the MoE methods, I-fold of the Mid and IV for the LMF.

Table 3. The comparison of four fusion methods performed on the Penn Haptic Texture Toolkit (HaTT) dataset.

	I	II	III	IV	V	Mean
Late	79.5	80.8	79.4	78.3	79.5	79.5 ± 0.9
MoE	77.9	78.9	74.3	76.6	73.4	76.2 ± 2.3
Mid	78.9	75.4	79.8	78.3	76.6	77.8 ± 1.8
LMF	78.1	80.9	78.9	78.3	79.5	79.1 ± 1.1

Table 4. The comparison of four fusion methods performed on the PHAC-2 dataset. All values represent the Area Under a Curve (AUC)-Receiver Operating Characteristic (ROC) performance metric.

	I	II	III	IV	V	Mean
Late	0.923	0.924	0.922	0.923	0.925	0.923 ± 0.001
MoE	0.923	0.919	0.919	0.923	0.927	0.922 ± 0.003
Mid	0.929	0.922	0.922	0.927	0.925	0.925 ± 0.003
LMF	0.896	0.898	0.902	0.908	0.900	0.901 ± 0.005

4.2. Data Degradation Robustness

In the following section we present the results gathered from experiments regarding the robustness of selected methods against a variety of input data degradation scenarios. The research carried out brought very important conclusions on the capabilities of each fusion method to translate knowledge from one modality to another and revealed that in most cases there exists a phenomenon, which we called a leading modality. Namely, for each dataset there was a modality, the leading one, which regardless of the fusion method used is crucial for obtaining good results. To make this dependency visible, we presented our results in the form of heat-maps (see Figures 4–6).

In heat-maps, there were presented changes in a performance caused by decreasing the quality of one or more input modalities. We tested fusion methods against scenarios described below (a–e) and each row in heat-maps corresponds to one of the scenarios:

(a) N—a normal noise N added to selected modalities with a 0 mean and 0.7 standard deviation;
(b) U—a uniform noise U added to selected modalities that varies in the range (-0.5 to 0.5);
(c) 0—setting zeros in place of selected modalities, what simulated a deactivated/broken sensor;
(d) RN—replacing selected modalities with normal noise N;
(e) RU—replacing selected modalities with normal noise U.

Each heat-map column was annotated by a number that specify affected modalities (e.g., by the added uniform noise). For each dataset we tested fusion methods using three input modalities numbered as follows:

(a) BiGS—1: gripper positions, 2: gripper spatial orientations, 3: 3-axis force;
(b) HaTT—1: normal force, 2: squashed acceleration, 3: squashed velocity;
(c) PHAC-2—1: images, 2: raw electrodes from the 1st sensor, 3: raw electrodes from the 2nd sensor.

At first, we tested selected fusion methods on the BiGS dataset and visualized the results on heat-maps in Figure 4. The use of heat-maps enabled one to easily inspect the knowledge alignment and translation properties of each fusion method. To perform these tests, the best performing models from Table 2 (marked in blue) were used.

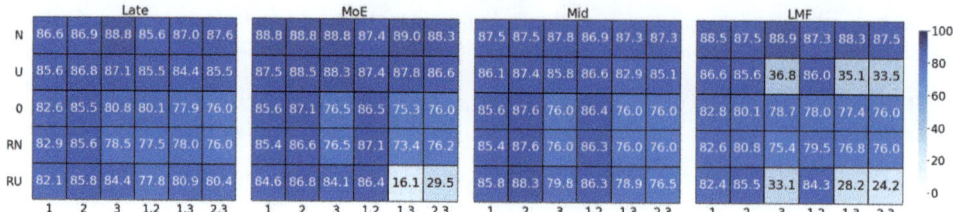

Figure 4. Results achieved by chosen models from the first stage of experiments on degraded data from the BiGS dataset. In heat-maps, there was presented a classification accuracy expressed in [%]. Rows correspond to different degradation scenarios, while columns are annotated by the indexes of affected modalities.

In Figure 5 we presented heat-maps generated for tests on the HaTT dataset. The influence of each modality on the final prediction is clearly visible and not every method is able to manage data degradation. Moreover, the 2nd modality (acceleration) appeared to be the leading one what resulted in a significant drop when it was noisy or faded. On the other hand, removing other modalities from the input data stream did not affect the final accuracy.

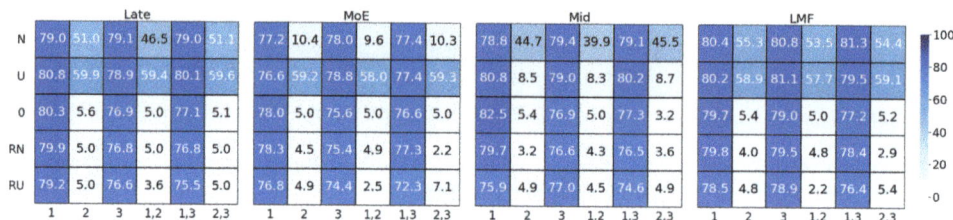

Figure 5. Accuracy [%] of a texture classification achieved while testing different fusion methods on degraded data from the HaTT dataset. Leading modality played a decisive dominant role, which resulted in a decreased quality in case of its degradation.

The AUC-ROC metric for the multi-label classification of haptic adjectives was reported in Figure 6. Similarly as in the experiments on the HaTT dataset, the leading modality is also visible. However, its correlations with other modalities played an even more important role in the final performance of methods. Inspecting heat-maps one can observe that the most meaningful correlations for predictions are between images (1st) and raw electrodes signals (2nd and 3rd). On the other hand. Noised interactions between both electrodes' time series only slightly influenced the classification performance.

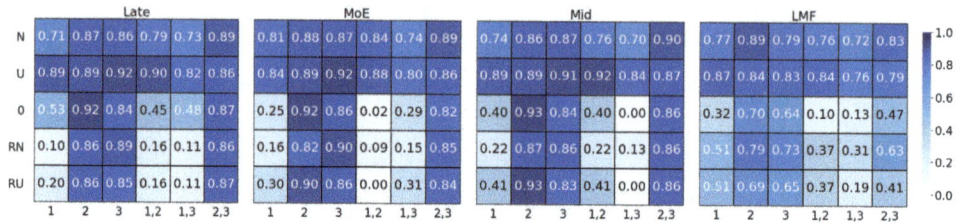

Figure 6. The AUC-ROC reported for the multi-label classification task using the PHAC-2 dataset. The leading modality is visible, but the correlations between modalities also affect predictions.

4.3. Data Augmentation vs. Leading Modality

Generated heat-maps from the previous stage of experiments revealed that in every dataset there exist one leading modality that had the biggest impact on the prediction. To verify whether the degradation of the leading modality was a determining factor for the decreased performance of fusion methods we conducted more experiments using the data augmentation technique and removing a leading modality from the input data stream. In the following set of experiments we noised/faded the 33% of randomly selected training samples of a leading modalities in the training dataset. In the degraded part of the dataset, a half of samples were zeroed, while others were noised with the normal noise with the 0 mean and standard deviation set to 0.7. This value should be related the fact that standard deviation of the data was confined to one through data standardization described in Section 3.1. Again, to perform the experiments we chose models from the Section 4.1 marked with a blue color and re-trained them on the same folds as they were trained originally. As in previous experiments, all of the methods were tested again separate test subsets. Results were reported in Tables 5–7.

First of all, we re-trained fusion models on the augmented training dataset and tested them on the same versions of test datasets as in the Section 4.1. We did that to verify the impact of the data degradation on the performance on test sets without it. An accuracy [%] and AUC-ROC values [-] were presented in Table 5.

Table 5. Outcomes of multi-modal fusion methods obtained for models trained on datasets containing noised/zeroed inputs from the leading modality, but tested on the dataset without such samples.

	Late	MoE	Mid	LMF
BiGS [%]	88.26	88.89	88.64	89.39
HaTT [%]	78.15	75.8	75.1	76.95
PHAC-2 [-]	0.92	0.93	0.93	0.91

Secondly, we evaluated the performance of each fusion method on test datasets with the same proportion of the noised samples of the leading modality without zeroed samples. Results obtained during that trial were reported in Table 6.

Table 6. Results obtained for models trained on datasets containing noised/zeroed inputs from the leading modality and a noised leading modality channel during tests.

	Late	MoE	Mid	LMF
BiGS [%]	88.14	88.64	88.51	89.26
HaTT [%]	65.85	69.95	65.3	68.3
PHAC-2 [-]	0.88	0.86	0.88	0.84

Finally, the influence of zeroed leading modalities on fusion methods was assessed during the last stage of experiments. Table 7 contains accuracy and AUC-ROC metric results gathered on the test dataset, where the leading modality was switched-off.

Table 7. Results obtained for models trained on datasets containing noised/zeroed inputs from the leading modality and a zeroed leading modality during tests.

	Late	MoE	Mid	LMF
BiGS [%]	85.77	88.76	86.27	89.01
HaTT [%]	54.85	57.05	53.95	53.25
PHAC-2 [-]	0.23	0.24	0.21	0.61

5. Discussion

The discussion of our experiments was divided into two parts—the discussion on the performance of each method and the analysis of their properties. In our work, results revealed that methods most widely spread in robotics can successfully deal with multi-modal data, however, they exhibit a fragility to missing/noised modalities, which resulted in decreased performance in such situations. That phenomenon might be considered as the main drawback of data-driven approaches to multi-modal fusion. However, as neural networks are included among data-driven methods, we were able to reduce to some extent the adverse impact of input quality degradation using the data augmentation technique.

5.1. Comparison of Fusion Methods

We compared the performance of four multi-modal fusion methods on three datasets including different time series (accelerations, velocities, forces), spatial transformations, and images. The important thing to note is that our target was to examine methods on the possibly largest set of homo and heterogeneous signals, hence we used different sets of modalities from each dataset and no modality was repeated between datasets even though, e.g., in the BiGS and PHAC-2, there were used the same tactile sensors—BioTacs.

The mean accuracy of all methods tested on the BiGS dataset was 88% and the differences between them were insignificant. Nevertheless, the most efficient fusion method in the grasp classification task was the LMF due to a smaller standard deviation among folds than the second method—the Mid fusion. The HaTT tests exhibited a slight increase of a mean results variance among different methods and standard deviations among folds comparing to BiGS results. In the texture recognition classification based on haptic signals, the best method turned out to be Late fusion, achieving a mean test accuracy of 79.5% between folds and additionally the smallest standard deviation equal to 0.9%. In the multi-label classification of haptic adjectives based on visual and haptic data the Late, MoE, and the Mid fusion methods were extremely close to each other in terms of a mean AUC-ROC metric, achieving a result of 0.92. The LMF turned out to be marginally below the performance represented by other methods.

On all tested datasets, the results from 5-fold cross-validation and tests on separate subsets showed that data used in experiments was consistent and there were no significant outliers among folds. That finding made it possible to carry out reliable experiments and ensure a fair comparison. Mean values of performance metrics among different fusion methods may suggest that the type of data fusion affects the performance of the model only to a very small extent. When all modalities are available and free of any noise it seems that more important was a reliable data preparation (e.g., ensuring a balanced distribution between classes in the train and test subsets, as well as between folds) for the training procedure than the fusion algorithm itself. In that section, apart from the comparison of different methods, we trained and chose the best-performing models for the next experiments. Chosen models were marked with blue color in Tables 2–4. All of the tested neural networks were trained in an end-to-end manner and performed relatively well, exhibiting a great capacity to learn from large (BiGS, HaTT) and small (PHAC-2) datasets. Further tests were conducted towards an assessment of the impact of each modality on the final result, and verifying the robustness of each method against input data degradation.

5.2. Data Degradation Robustness

All experiments from Sections 4.1–4.3 revealed the existence of a leading modality, which means that in our experiments there was always a one modality that played a dominant role for the discrimination between classes.

In Figure 4, one can observe that, for BiGS dataset, MoE and LMF fusion methods exhibited significantly decreased performance when the modality no. 3, which was a 3-axis force signal in combination with other modalities, was replaced by the uniform noise. Hence, we consider this signal as a leading modality in the BiGS dataset. As it can be observed, the LMF and MoE were also sensitive to the uniform noise added that affected the force signal. However, the described phenomenon did not occur for other fusion methods—the Late and Mid. They appeared to be relatively robust for data degradation scenarios, exhibiting no more than 10% of a drop in the accuracy when the leading modality was noised, zeroed or totally replaced by noise. In a case of MoE the drop in the quality of discrimination between classes to the 16.1% and 29.5% would be caused by the fact that during the training, the gating network was trained to increase the importance of a leading modality for the prediction because it played a dominant role in provided data. Thus, when that signal, together with any of the other modality, were replaced by the noise, too much emphasis was put on the correlation of these two signals, what resulted in prediction mistakes. However, MoE exhibited robustness for any other data degradation scenarios. It appeared to be sensitive mainly to the correlations between the force signal and two other modalities—hand positions and orientations. Henceforth, in the MoE, the interactions between the modalities were essential for the grasp outcome classification, not the leading modality itself. Similar results were achieved while testing the LMF method—when the force signal was replaced by the uniform noise, the method was more likely to make mistakes, thus the interactions between modalities were meaningless. Additionally, the LMF was sensitive to not only scenarios with replacing the leading modality with noise and its combinations with other modalities, but also for the noise added to the signals. This could be observed in the LMF heat-map in Figure 4 looking at the results (36.8%, 35.1%, and 33.5%) in the U-row. The described phenomenon might be caused due to the fact that LMF is a tensor-based method that highly relies on outer products between uni-modal representations inside networks, thus the highest emphasis was put on inter-modality interactions. When a finding of these interactions was difficult/not possible, the LMF struggled to find a correct prediction for the grasp outcome evaluation task. It can be also observed that the type of noise introduced to the input data played a significant role for the prediction performance because presented findings were not observed for a normal noise. To explain that effect, we speculate that during the training phase, in signals there was already some noise present similar to the normal level, which resulted in a higher robustness for such a data degradation. The achieved robustness was truly substantial and it appears that a balance between the importance of modalities was paramount. Nevertheless, verifying that relevance is very challenging and involves a great number of experiments.

Another dataset involved in experiments was the HaTT, and results gathered during that trial were reported in Figure 5. Contrary to the BiGS dataset, by inspecting heat-maps one can observe that the 2nd modality (an acceleration) caused a significant drop of the accuracy for all tested data degradation scenarios and fusion methods. Hence, we consider an acceleration to be a leading modality from proposed set of input modalities. In the task of texture classification based on haptic signals, all methods exhibited a similar performance and sensitivity on different disturbances. A lack of a legitimate acceleration signal (zeroing or replacing with a noise) always caused a decreased performance to the level of 5% for all tested methods. This means that in the proposed set of modalities, the domination of an acceleration was tremendous, and the rest of the signals did not provide meaningful information about the process under investigation. The late fusion and LMF heat-maps evaluation gave a similar results for all data degradation scenarios, but the MoE and Mid fusion differs in terms of managing the added noise—MoE exhibited sensitivity to an appearance of the noise component in the leading modality, but the Mid was fragile for the same phenomenon but with the uniform noise added.

The results of the multi-label classification of haptic adjectives performed on the PHAC-2 dataset were shown in Figure 6. The biggest drop in the performance metric was reported for columns missing the 1st modality—an image, which was considered as a leading modality. In the MoE, the fact that the lack of images was able to cause a total failure of the classifier achieving the result of 0 (which means that the modal was always wrong) again indicates that gating network during a training put too much emphasis on the dominant modality. Taking into account that every result below 0.5 level means that the classifier is more often wrong than right on average. The Late, MoE and Mid fusion methods behaved similarly across all scenarios—the performance without a leading modality was significantly decreased. The LMF performed slightly different, achieving relatively good results when an input image was replaced by a noise what can be seen in the first column of the LMF heat-map. However, it performed worst in case of scenarios where other modalities were replaced by a noise/zeroed.

Although, it should be noted, that it does not achieved 0 AUC-ROC as it happened in case of MoE and Mid methods. Additionally, sometimes one can observe an improvement in the classification performance achieved when one modality was noised/faded. Such a phenomenon was reported, e.g., for the Mid fusion when the 2nd modality (raw electrode signal) was zeroed or replaced by a uniform noise. Comparing to Table 4 the improvement was 3%.

5.3. Data Augmentation vs. Leading Modality

In the last stage of experiments we re-trained models from the Section 4.1. We did that on the same folds as in the original experiment to ensure a fair comparison and provide results comparable with performed comparison of fusion methods.

First of all, we augmented the training dataset by adding normal noise $\mathcal{N}(0, 0.7)$ or switching-off the leading modality in 33% of the samples. Then, we re-trained selected models on corresponding folds and reported a mean accuracy and AUC-ROC metric of the classification for test datasets in Table 5. By measuring the influence of the data augmentation we established a point of reference for further experiments, as such augmentation could decrease the performance. As we can observe, we obtained similar results as in Tables 2–4, which means that this partial degradation does not affect the performance in nominal conditions significantly. In Tables 6 and 7 we reported the results for the degraded test set. In Table 5 the results from tests for noised leading modality were presented, whereas in Table 7 those for zeroed modality were presented.

As it can be observed in both tables, the data augmentation procedure increased robustness on noised and missing modalities entirely for the BiGS dataset, thus all methods gave similar results as during the tests on the data without any degradation applied on an leading modality. We believe that a proposed data augmentation procedure is sufficient to ensure a robustness on noised/missing samples for the proposed set of input modalities.

However, the above statement is not always true, which is clearly visible in results obtained for the HaTT dataset, when the mean decrease of accuracy was from 6% to 13% when comparing Table 5 to Table 6 and even larger from 18% to 24% between Tables 5 and 7. The results proved the same conclusions as before—the leading modality in the HaTT dataset possessed so much information meaningful for the discrimination between textures and other modalities played only a supporting role for that task. Nevertheless, using data augmentation still brought a significant improvement in results comparing to data degradation scenarios showed in Table 5. In both tested variants, the best performing method turned out to be the MoE, which indicates that the gating network learned to more efficiently refuse a predictions based on a degraded leading modality.

In the multi-label classification task on the PHAC-2 the Late, Mid and MoE methods failed to properly assign haptic adjectives when the vision was missing. However, the LMF method apparently was able to find intra- and inter-modality interactions that led to the surprisingly good result of 0.61 AUC-ROC metric. It indicates that the LMF was the only method that was able to actually assign the haptic adjective properly more often than make a mistake on average. In tests involving noise-only samples, all methods achieved similar result and the performance metric dropped only by 4–6%.

6. Conclusions

In our work we compared four multi-modal fusion methods that could be regarded as state-of-the-art. We assessed their performance in the three tasks—a prediction of a grasp outcome, a texture recognition and multi-label classification of haptic adjectives. Then, selected methods were verified in the variety of possible scenarios of input data degradation that might occur in real life, e.g., a sensor turn-off or a measurement noise. Finally, we measured the influence of data augmentation technique on the predictive capabilities of tested methods and again evaluated their robustness on noise added to the leading modality and its zeroing. We hope that the findings contained in our paper will make researchers realize that State-of-the-Art fusion methods are prone to over-fit to specific modalities, so-called leading modalities, and are rather susceptible to the noise as well as sensor failures. Thus, in order to build reliable autonomous systems, we have to focus more on the robustness of our data fusion methods. Due to that, all our code and data used in experiments were made available as open-source.

Author Contributions: M.B. and P.K. conceived and designed the experiments; M.B. performed the experiments; M.B. and P.K. analyzed the data; M.B., P.K. and K.W. wrote the paper. All authors have read and agreed to the published version of the manuscript.

Funding: This work was supported by grant No. LIDER/3/0183/L-7/15/NCBR/2016 funded by The National Centre for Research and Development (Poland). M.B and K.W. were partially supported by the PUT Faculty of Control, Robotics and Electrical Engineering grant SBAD/0207 in year 2019.

Conflicts of Interest: The authors declare no conflict of interest.

References

1. McGurk, H.; MacDonald, J. Hearing lips and seeing voices. *Nature* **1976**, *264*, 746–748. [CrossRef] [PubMed]
2. Baltrusaitis, T.; Ahuja, C.; Morency, L.P. Multimodal Machine Learning: A Survey and Taxonomy. *IEEE Trans. Pattern Anal. Mach. Intell.* **2019**, *41*, 423–443. [CrossRef] [PubMed]
3. Yuan, W.; Dong, S.; Adelson, E. GelSight: High-Resolution Robot Tactile Sensors for Estimating Geometry and Force. *Sensors* **2017**, *17*, 2762. [CrossRef] [PubMed]
4. Izatt, G.; Mirano, G.; Adelson, E.; Tedrake, R. Tracking objects with point clouds from vision and touch. In Proceedings of the 2017 IEEE International Conference on Robotics and Automation (ICRA), Marina Bay Sands, Singapore, 29 May–3 June 2017; pp. 4000–4007. [CrossRef]
5. Wettels, N.; Santos, V.; Johansson, R.; Loeb, G. Biomimetic Tactile Sensor Array. *Adv. Robot.* **2008**, *22*, 829–849. [CrossRef]
6. Fazeli, N.; Oller, M.; Wu, J.; Wu, Z.; Tenenbaum, J.; Rodriguez, A. See, feel, act: Hierarchical learning for complex manipulation skills with multisensory fusion. *Sci. Robot.* **2019**, *4*, eaav3123. [CrossRef]
7. Lee, M.A.; Zhu, Y.; Srinivasan, K.; Shah, P.; Savarese, S.; Fei-Fei, L.; Garg, A.; Bohg, J. Making Sense of Vision and Touch: Self-Supervised Learning of Multimodal Representations for Contact-Rich Tasks. In Proceedings of the 2019 International Conference on Robotics and Automation (ICRA), Montreal, QC, Canada, 20–24 May 2019; pp. 8943–8950.
8. Mees, O.; Eitel, A.; Burgard, W. Choosing Smartly: Adaptive Multimodal Fusion for Object Detection in Changing Environments. In Proceedings of the IEEE/RSJ International Conference on Intelligent Robots and Systems (IROS), Daejeon, Korea, 9–14 October 2016.
9. Valada, A.; Vertens, J.; Dhall, A.; Burgard, W. AdapNet: Adaptive semantic segmentation in adverse environmental conditions. In Proceedings of the 2017 IEEE International Conference on Robotics and Automation (ICRA), Marina Bay Sands, Singapore, 29 May–3 June 2017; pp. 4644–4651.
10. Hung, C.; Nieto, J.; Taylor, Z.; Underwood, J.; Sukkarieh, S. Orchard fruit segmentation using multi-spectral feature learning. In Proceedings of the 2013 IEEE/RSJ International Conf. on Intelligent Robots and Systems, Tokyo, Japan, 3–7 November 2013; pp. 5314–5320.
11. Ilonen, J.; Bohg, J.; Kyrki, V. Fusing visual and tactile sensing for 3-D object reconstruction while grasping. In Proceedings of the 2013 IEEE International Conference on Robotics and Automation, Karlsruhe, Germany, 6–10 May 2013; pp. 3547–3554. [CrossRef]
12. Gao, Y.; Hendricks, L.A.; Kuchenbecker, K.J.; Darrell, T. Deep learning for tactile understanding from visual and haptic data. In Proceedings of the 2016 IEEE International Conference on Robotics and Automation (ICRA), Stockholm, Sweden, 16–21 May 2016; pp. 536–543. [CrossRef]
13. Liu, Z.; Shen, Y.; Lakshminarasimhan, V.B.; Liang, P.P.; Zadeh, A.; Morency, L.P. Efficient Low-rank Multimodal Fusion with Modality-Specific Factors. *arXiv* **2018**, arXiv:1806.00064.
14. Ramachandram, D.; Taylor, G.W. Deep Multimodal Learning: A Survey on Recent Advances and Trends. *IEEE Signal Process. Mag.* **2017**, *34*, 96–108. [CrossRef]
15. Castanedo, F. A Review of Data Fusion Techniques. *Sci. World J.* **2013**, *2013*, 704504. [CrossRef] [PubMed]
16. Valada, A.; Oliveira, G.; Brox, T.; Burgard, W. Deep Multispectral Semantic Scene Understanding of Forested Environments using Multimodal Fusion. In *International Symposium on Experimental Robotics (ISER 2016), Proceedings of the 2016 International Symposium on Experimental Robotics, Tokyo, Japan, 3–6 October 2016*; Springer: Cham, Switzerland, 2016.

17. Kopuklu, O.; Kose, N.; Rigoll, G. Motion Fused Frames: Data Level Fusion Strategy for Hand Gesture Recognition. In Proceedings of the The IEEE Conference on Computer Vision and Pattern Recognition (CVPR) Workshops, Salt Lake City, UT, USA, 18–22 June 2018.
18. Walas, K.; Nowicki, M.; Ferstl, D.; Skrzypczyński, P. Depth Data Fusion for Simultaneous Localization and Mapping—RGB-DD SLAM. In Proceedings of the 2016 IEEE International Conference on Multisensor Fusion and Integration for Intelligent Systems (MFI 2016), Baden-Baden, Germany, 19–21 September 2016; pp. 9–14.
19. Liu, H.; Li, F.; Xu, X.; Sun, F. Multi-modal local receptive field extreme learning machine for object recognition. *Neurocomputing* **2018**, *277*, 4–11. [CrossRef]
20. Xiong, Z.; Yuan, Y.; Wang, Q. RGB-D Scene Recognition via Spatial-Related Multi-Modal Feature Learning. *IEEE Access* **2019**, *7*, 106739–106747. [CrossRef]
21. Sebastian, B.; Ren, H.; Ben-Tzvi, P. Neural Network Based Heterogeneous Sensor Fusion for Robot Motion Planning. In Proceedings of the 2019 IEEE/RSJ International Conference on Intelligent Robots and Systems (IROS), Macau, China, 3–8 November 2019; pp. 2899–2904.
22. Choi, J.; Lee, J.S. EmbraceNet: A robust deep learning architecture for multimodal classification. *Inf. Fusion* **2019**, *51*, 259–270. [CrossRef]
23. Zadeh, A.; Chen, M.; Poria, S.; Cambria, E.; Morency, L.P. Tensor Fusion Network for Multimodal Sentiment Analysis. In Proceedings of the 2017 Conference on Empirical Methods in Natural Language Processing, Copenhagen, Denmark, 9–11 September 2017; Association for Computational Linguistics: Copenhagen, Denmark, 2017; pp. 1103–1114. [CrossRef]
24. Hou, M.; Tang, J.; Zhang, J.; Kong, W.; Zhao, Q. Deep Multimodal Multilinear Fusion with High-order Polynomial Pooling. In *Advances in Neural Information Processing Systems 32*; Wallach, H., Larochelle, H., Beygelzimer, A., d' Alché-Buc, F., Fox, E., Garnett, R., Eds.; Curran Associates, Inc.: Red Hook, NY, USA, 2019; pp. 12136–12145.
25. Kittler, J.; Hatef, M.; Duin, R.P.W.; Matas, J. On Combining Classifiers. *IEEE Trans. Pattern Anal. Mach. Intell.* **1998**, *20*, 226–239. [CrossRef]
26. Zhang, R.; Candra, S.A.; Vetter, K.; Zakhor, A. Sensor fusion for semantic segmentation of urban scenes. In Proceedings of the 2015 IEEE International Conference on Robotics and Automation (ICRA), Seattle, WA, USA, 26–30 May 2015; pp. 1850–1857.
27. Eitel, A.; Springenberg, J.T.; Spinello, L.; Riedmiller, M.; Burgard, W. Multimodal deep learning for robust RGB-D object recognition. In Proceedings of the 2015 IEEE/RSJ International Conference on Intelligent Robots and Systems (IROS), Hamburg, Germany, 28 September–2 October 2015; pp. 681–687.
28. García, G.M.; Potapova, E.; Werner, T.; Zillich, M.; Vincze, M.; Frintrop, S. Saliency-based object discovery on RGB-D data with a late-fusion approach. In Proceedings of the 2015 IEEE International Conference on Robotics and Automation (ICRA), Seattle, WA, USA, 26–30 May 2015; pp. 1866–1873.
29. Tian, J.; Cheung, W.; Glaser, N.; Liu, Y.C.; Kira, Z. UNO: Uncertainty-Aware Noisy-Or Multimodal Fusion for Unanticipated Input Degradation. Available online: https://nikosuenderhauf.github.io/roboticvisionchallenges/assets/papers/IROS19/tian.pdf (accessed on 10 April 2020).
30. Walas, K. Terrain Classification and Negotiation with a Walking Robot. *J. Intell. Robot. Syst.* **2015**, *78*, 401–423. [CrossRef]
31. Kim, J.; Koh, J.; Kim, Y.; Choi, J.; Hwang, Y.; Choi, J.W. Robust Deep Multi-modal Learning Based on Gated Information Fusion Network. In *Computer Vision—ACCV 2018, Proceedings of the 14th Asian Conference on Computer Vision, Perth, Australia, 2–6 December 2018*; Jawahar, C., Li, H., Mori, G., Schindler, K., Eds.; Springer International Publishing: Cham, Switzerland, 2019; pp. 90–106.
32. Kim, T.; Ghosh, J. On Single Source Robustness in Deep Fusion Models. In *Advances in Neural Information Processing Systems 32, Proceedings of the Neural Information Processing Systems 2019, Vancouver, BC, Canada, 8–14 December 2018*; Wallach, H., Larochelle, H., Beygelzimer, A., d' Alché-Buc, F., Fox, E., Garnett, R., Eds.; Curran Associates, Inc.: Red Hook, NY, USA, 2019; pp. 4814–4825.
33. Bijelic, M.; Muench, C.; Ritter, W.; Kalnishkan, Y.; Dietmayer, K. Robustness Against Unknown Noise for Raw Data Fusing Neural Networks. In Proceedings of the 2018 21st International Conference on Intelligent Transportation Systems (ITSC), Maui, HI, USA, 4–7 November 2018; pp. 2177–2184.
34. Patel, N.; Choromanska, A.; Krishnamurthy, P.; Khorrami, F. A deep learning gated architecture for UGV navigation robust to sensor failures. *Robot. Auton. Syst.* **2019**, *116*, 80–97. [CrossRef]

35. Sechidis, K.; Tsoumakas, G.; Vlahavas, I. On the stratification of multi-label data. In *Machine Learning and Knowledge Discovery in Databases, Proceedings of the European Conference, ECML PKDD 2017, Skopje, Macedonia, 18–22 September, 2017*; Springer: Berlin/Heidelberg, Germany, **2011**; pp. 145–158.
36. Szymański, P.; Kajdanowicz, T. A Network Perspective on Stratification of Multi-Label Data. In Proceedings of the First International Workshop on Learning with Imbalanced Domains: Theory and Applications, Skopje, Macedonia, 22 September 2017 ; Volume 74, pp. 22–35.
37. Chebotar, Y.; Hausman, K.; Su, Z.; Molchanov, A.; Kroemer, O.; Sukhatme, G.; Schaal, S. BiGS: BioTac Grasp Stability Dataset. In *ICRA 2016 Workshop on Grasping and Manipulation Datasets, Proceedings of ICRA 2016—IEEE International Conference on Robotics and Automation, Stockholm, Sweden, 16–20 May 2016*.
38. Culbertson, H.; Lopez Delgado, J.J.; Kuchenbecker, K.J. The Penn Haptic Texture Toolkit for Modeling, Rendering, and Evaluating Haptic Virtual Textures. Available online: https://repository.upenn.edu/cgi/viewcontent.cgi?article=1311&context=meam_papers (accessed on 10 April 2020).
39. Slobodenyuk, N.; Jraissati, Y.; Kanso, A.; Ghanem, L.; Elhajj, I. Cross-Modal Associations Between Color and Haptics. *Attention Percept. Psychophys.* **2015**, *77*, 1379–1395. [CrossRef] [PubMed]
40. Landin, N.; Romano, J.M.; McMahan, W.; Kuchenbecker, K.J. Dimensional Reduction of High-Frequency Accelerations for Haptic Rendering. In *Haptics: Generating and Perceiving Tangible Sensations, Proceedings of the International Conference, EuroHaptics 2010, Amsterdam, The Netherlands, 8–10 July 2010*; Kappers, A.M.L., van Erp, J.B.F., Bergmann Tiest, W.M., van der Helm, F.C.T., Eds.; Springer: Berlin/Heidelberg, Germany, 2010; pp. 79–86.
41. Bell, J.; Bolanowski, S.; Holmes, M.H. The structure and function of pacinian corpuscles: A review. *Prog. Neurobiol.* **1994**, *42*, 79–128. [CrossRef]
42. Chu, V.; McMahon, I.; Riano, L.; McDonald, C.G.; He, Q.; Perez-Tejada, J.M.; Arrigo, M.; Darrell, T.; Kuchenbecker, K.J. Robotic learning of haptic adjectives through physical interaction. *Robot. Auton. Syst.* **2015**, *63*, 279–292. [CrossRef]

© 2020 by the authors. Licensee MDPI, Basel, Switzerland. This article is an open access article distributed under the terms and conditions of the Creative Commons Attribution (CC BY) license (http://creativecommons.org/licenses/by/4.0/).

MDPI
St. Alban-Anlage 66
4052 Basel
Switzerland
Tel. +41 61 683 77 34
Fax +41 61 302 89 18
www.mdpi.com

Electronics Editorial Office
E-mail: electronics@mdpi.com
www.mdpi.com/journal/electronics